T0292010

Power System Energy Storage Technologies

The Power Generation Series

Paul Breeze—Coal-Fired Generation, ISBN 13: 9780128040065
Paul Breeze—Gas-Turbine Fired Generation, ISBN 13: 9780128040058
Paul Breeze—Solar Power Generation, ISBN 13: 9780128040041
Paul Breeze—Wind Power Generation, ISBN 13: 9780128040386
Paul Breeze—Fuel Cells, ISBN 13: 9780081010396
Paul Breeze—Energy from Waste, ISBN 13: 9780081010426
Paul Breeze—Nuclear Power, ISBN 13: 9780081010433
Paul Breeze—Electricity Generation and the Environment,
ISBN 13: 9780081010440

Power System Energy Storage Technologies

Paul Breeze

ACADEMIC PRESS

An imprint of Elsevier

Academic Press is an imprint of Elsevier
125 London Wall, London EC2Y 5AS, United Kingdom
525 B Street, Suite 1800, San Diego, CA 92101-4495, United States
50 Hampshire Street, 5th Floor, Cambridge, MA 02139, United States
The Boulevard, Langford Lane, Kidlington, Oxford OX5 1GB, United Kingdom

Notices
Knowledge and best practice in this field are constantly changing. As new research and experience broaden our
understanding, changes in research methods, professional practices, or medical treatment may become
necessary.

Practitioners and researchers must always rely on their own experience and knowledge in evaluating and using
any information, methods, compounds, or experiments described herein. In using such information or methods
they should be mindful of their own safety and the safety of others, including parties for whom they have a
professional responsibility.

To the fullest extent of the law, neither the Publisher nor the authors, contributors, or editors, assume any
liability for any injury and/or damage to persons or property as a matter of products liability, negligence or
otherwise, or from any use or operation of any methods, products, instructions, or ideas contained in the
material herein.

British Library Cataloguing-in-Publication Data
A catalogue record for this book is available from the British Library

Library of Congress Cataloging-in-Publication Data
A catalog record for this book is available from the Library of Congress

ISBN: 978-0-12-812902-9

For Information on all Academic Press publications
visit our website at https://www.elsevier.com/books-and-journals

Working together
to grow libraries in
developing countries

www.elsevier.com • www.bookaid.org

Publisher: Joe Hayton
Acquisition Editor: Lisa Reading
Editorial Project Manager: Mariana L. Kuhl
Production Project Manager: Vijayaraj Purushothaman
Cover Designer: MPS

Typeset by MPS Limited, Chennai, India

CONTENTS

CHAPTER 1

An Introduction to Energy Storage Technologies

Energy storage technologies comprise a range of different systems that can take up and store electrical energy, holding it securely and making it available for delivery at a later time. These systems vary in the amount of energy they can hold and in the time over which they can store the energy without significant loss. They can operate at the grid level, absorbing surplus energy from the grid and then delivering it back when grid demand rises, or they can operate at the individual consumer level, balancing local supply and demand. In either case the storage system will reduce peaks and troughs in demand from the grid and help stabilize the supply.

Electrical storage technologies are also capable of offering a range of vital grid services, particularly in the support of grids that absorb large volumes of renewable energy from intermittent sources such as wind and solar power. This makes them increasingly attractive as the use of these renewable technologies grows. In the past, the availability of energy storage technology has been limited, often because capacity has been considered too expensive to build.

The earliest type of energy storage technology for grid use was pumped storage hydropower and this continues to be the most important in terms of absolute capacity. New technologies including batteries and flywheels provide smaller scale storage units but with the advantage of faster response. Advances, particularly among batteries, are making these technologies cheaper and more widely accessible and this could lead to a transformation over the next two to three decades in the way grids operate.

AN ENERGY STORAGE OVERVIEW

While the storage of electrical energy is still relatively rare, other forms of energy storage play a vital part in the modern global economy. At a

Power System Energy Storage Technologies. DOI: https://doi.org/10.1016/B978-0-12-812902-9.00001-8

national level, oil and gas are regularly stored by both utilities and by governments while at a smaller scale petrol stations store gasoline and all cars carry a storage tank to provide them with the ability to travel a significant distance between refueling stops. Domestic storage of hot water is also usual in modern homes. Yet when it comes to electrical energy, storage on anything but a small scale, in batteries, is still uncommon.

Part of the reason for this is that storage of electricity, although it can be achieved in a number of ways, is far from straightforward. In most storage technologies, the electricity must be converted into some other form of energy before it can be stored. For example, in a battery it is converted into chemical energy while in a pumped storage hydropower plant the electrical energy is turned into the potential energy contained within an elevated mass of water. Energy conversion makes the storage process complex and the conversion itself is often inefficient. These and other factors help to make energy storage system costly.

In spite of such obstacles, large-scale energy storage plants have been built in many countries. By far the largest part of these, in terms of installed capacity is provided by pumped storage hydropower plants, often built to capture and store power from base-load nuclear power plants during off-peak periods. Many of these storage plants were built in the 1960s, 1970s, and 1980s. More recently there has been renewed interest in technologies such as pumped storage for grid support, particularly in European countries that are installing large capacities of renewable capacity such as wind and solar power. However the economics of energy storage often makes construction difficult to justify in liberalized electricity market. New tariffs that encourage energy arbitrage and grid support services may offer better economic incentives in the future.

While economics may not always favor their construction, energy storage plants offer significant benefits for the generation, distribution, and use of electric power. At the utility level, for example, a large energy storage facility can be used to store electricity generated during off-peak periods—typically overnight—and this energy can be delivered during peak periods of demand when the marginal cost of generating additional power can be several times the off-peak cost. Energy arbitrage of this type is potentially a lucrative source of revenue for storage plant operators and is how most pumped storage plants operate.

At a smaller scale, energy storage plants can supply emergency backup, as well as other grid support features, helping to maintain grid stability. Small, fast-operating storage units can be employed in factories or offices to take over in case of a grid supply failure. Indeed, in a critical facility such as a computer server facility where an instantaneous response to loss of power is needed, a storage technology that can take over within the space of a single cycle of the grid supply may be the only way to ensure complete reliability.

Energy storage also has an important role to play in the efficient use of electricity from renewable energy. Many renewable sources of energy such as solar, wind, and tidal energy are intermittent and so incapable of supplying electrical power continuously. Combining a renewable energy source with some form of energy storage helps remove this uncertainty and increases the value of the electricity generated. It also allows all the renewable energy available from these sources to be used. Today the shedding of excess renewable power when demand does not exist for it, or when the grid cannot cope with it, is becoming common on some grid systems with high renewable capacity.

While there are many types of electrical energy storage system, pumped storage hydropower plants account for virtually all grid storage capacity available today with perhaps 183 GW of generating capacity in operation, based on estimates by the US Department of Energy. This was effectively the only large-scale energy storage technology available until the late 1970s but in the past 30–40 years interest has been stimulated in a range of other technologies. These vary in size so that some are suitable for transmission system level storage while others are more suited to the distribution grid or even for small microgrids. They include a range of battery storage systems, compressed air-energy storage (CAES), large storage capacitors and flywheels, superconducting magnetic energy storage, and systems designed to generate hydrogen as an energy storage medium. The widespread adoption of electric vehicles that use battery energy storage could potentially offer a major new means of storing grid electricity too.

If deployed widely, these technologies could potentially transform the way the grid-based delivery of electrical energy is managed by eliminating the need for expensive peaking power plants while at the same time fully integrating the range of renewable generation technologies now available. This would, in turn, eliminate the vulnerability of electricity production to the economic vagaries of the global fossil fuel markets, creating more stable economic conditions everywhere. There is no consensus on how much storage capacity would be required to achieve this on a mature national grid but it could be equivalent to around 10%–15% of the available generating capacity. In principle, with much larger grid storage capacity, the use of energy storage might eliminate the need for fossil fuel power plants altogether, and with them one of the largest global sources of greenhouse gas emissions.

In spite of the apparent advantages offered by energy storage, widespread adoption remains slow. Cost appears to be the main obstacle although developments are slowly bringing costs down. At the same time, the growth of distributed generation is offering new opportunities for small-scale energy storage facilities.

THE HISTORY OF ENERGY STORAGE SYSTEMS

The early history of electricity storage is based entirely on the development and use of pumped storage hydropower. Opinions seem to differ as to when the first pumped storage plant was built but it was probably in Switzerland, either toward the end of the 19th century or at the beginning of the 20th century. Italy also adopted the technology at an early stage. These first plants used a turbine to generate electricity from falling water and a pump to raise water back into an upper reservoir using surplus electricity. Interest in this type of storage technology grew slowing during the early decades of the 20th century. It was not until the middle of the century that more, and larger plants started to be built. This surge in capacity growth was linked to the construction of nuclear power plants.

Nuclear power started to become popular in the 1960s and the plants that were built operated most effectively if they ran continuously at full output. Changing the output level to match demand was not easy to achieve. However the plants were large and there was often insufficient demand for their power overnight. To compensate, energy storage plants based on pumped storage hydropower were built to absorb this surplus power and then deliver it to the grid during the following day. The generating capacities of individual nuclear power plants soon grew to 1000 MW or more and pumped storage plants of similar size began to be built in the United States, Europe, and Japan, countries and regions where the growth in nuclear power was greatest. This led to large additions to storage capacity from the 1960s until the middle of the 1980s when the growth in nuclear power waned.

Even by the 1980s there was little alternative to pumped storage hydropower for practical electricity storage. One other large-scale technology, CAES, was explored around this time and two plants were built, one in Germany and the second in the United States. However although the technology offered some promise, it never expanded.

The other obvious technology for electricity storage was the rechargeable battery. This was explored toward the end of the 1980s when several large battery storage facilities based on lead-acid batteries were built. These had limited success but cell degradation over time proved a handicap for the early designs. Other secondary battery types such as nickel cadmium began to be tested at the beginning of the 21st century and during the second decade of this century interest in more modern battery technologies such a lithium ion cell for large-scale storage has grown.

The 1980s also saw interest in some more exotic storage technologies such as superconducting magnetic energy storage. This has continued although the storage capacity based on this technology is limited. However it has found appeal in grid support. Alternatives such as flywheels and super-capacitors also attracted attention. Meanwhile the idea of a hydrogen economy stimulated interest in hydrogen as an energy storage medium that could be used both for power generation and for transportation applications.

Since the 1980s there have been several studies that have supported the idea that energy storage can render a grid more stable, more reliable, and more economical to operate. At the same time, starting in the late 1980s, the supply of electricity was in many regions transferred from the public sector to the private sector. This made securing the investment needed for the construction of energy storage facilities more difficult because it required justification in market terms rather than in strategic terms as would have been the case previously. There has been a renaissance in interest in electrical storage technologies in the 21st century, particularly in some European countries like Spain where there has been large investment in renewable technologies such as wind and solar power. However most grids still lack sufficient storage capacity to make renewable integration easy.

TYPES OF ENERGY STORAGE TECHNOLOGY

The storage of electrical energy relies in almost all cases on the conversion of the electrical energy into another form of energy that can be stored easily. Only two types of system can store electricity directly, capacitors that store static electrical charge and superconducting magnetic storage rings that store an electrical current in an electrical coil with zero electrical resistance.[1] Pumped storage hydropower plants store potential energy associated with an elevated mass of water. Batteries convert electricity into chemical energy, as does hydrogen storage. Flywheels use motors to convert the electrical power into rotating mechanical energy. There are also systems that store thermal energy, which can be converted back into electricity and some experimental systems that store the energy in the form of liquefied air. Each type has different characteristics.

Pumped storage hydropower: Pumped storage hydropower currently provides the largest amount of energy storage capacity, globally. Individual plants can be larger than plants of any other type and the largest are up to 3000 MW in generating capacity. The technology depends on having two reservoirs of water separated in height. Water from the top reservoir can be used to generate electricity as it falls to the lower reservoir while energy can be stored by pumping water from the lower to the upper reservoir. A pumped storage plant can be brought on line in a matter of seconds.

Compressed air-energy storage: CAES is an energy storage technology based around the gas turbine. In a standard gas turbine, air is compressed then mixed with a fuel that is ignited, creating a high temperature, high-pressure gas stream that drives a turbine. In the CAES plant air is compressed and stored in chamber using surplus electricity. This compressed air can then be released through a turbine as needed to generate electricity. Additional energy can be gained by mixing this compressed air with fuel and igniting it, as in a conventional gas turbine.

Battery energy storage: Electrochemical batteries have been used for more than a century to provide a portable supply of electrical power. They rely on particular chemical reactions that can be exploited to produce electrical energy as they proceed. For energy storage a

---

[1]The energy in this type of storage device is contained in the magnetic field associated with the circulating current.

rechargeable cell is needed and this requires a special, reversible chemical process. There are a range of modern cells of this type including the lead-acid battery, the nickel cadmium battery, and lithium hydride and lithium ion batteries. These batteries are all compact, self-contained units. There is another type of battery called a flow battery that uses external stores for the chemical reactants and the products of the power-producing reaction. These are much less compact but are flexible in terms of the amount of energy they can store.

Thermal energy storage: Thermal energy storage uses a variety of media to store heat energy. This stored energy can then be used for heating or, depending on the quality of the heat, it can be used to generate electrical power. Various storage mediums are available including high-pressure, high-temperature steam, molten salts, and solid heat storage materials. Thermal storage is used in some solar thermal power plants to extend their operating window beyond the period when the sun is shining. It is not widely used elsewhere for electricity production.

Flywheels: Flywheels are mechanical devices that store energy as rotational inertia. A modern energy storage flywheel is coupled with a high-performance motor-generator that can feed energy into and extract energy from the rotating system very quickly. The devices rotate at extremely high speeds and must be constructed of special materials that are able to withstand the centrifugal forces exerted on the rotor. A flywheel generally stores only a small amount of electrical power but can respond and release the energy very quickly.

Capacitors: Capacitors (often called super-capacitors) are electrical devices that can store electrostatic charge. Capacitors for energy storage use advanced materials and techniques to store very large amounts of charge. This can be released very quickly and the devices can react extremely rapidly to the demand for power. However the amount of energy they can supply is still relatively small compared with most other storage technologies.

Liquid air energy storage: When air is cooled cryogenically, the gases will eventually form a liquid—liquid air. This is highly compressed compared to the gaseous state. If the liquid is then allowed to warm and return to the gaseous state, it will provide a flow of high-pressure gas that can be used to drive a turbine and generate power. The technology can, in principle, utilize waste heat and cold from industrial processes as well as surplus electrical power to produce the liquefied air. The technique is experimental but its promoters claim it can use off the shelf components to provide an economical form of electricity storage and release.

Hydrogen storage: Hydrogen is a flammable gas that can be produced by the electrolysis of water. When it burns in oxygen or air, it produces only water, making it a very clean fuel. It can be used as a fuel in various power-generating systems as well as providing transportation fuel and could replace natural gas for domestic use too. Since the gas can easily be produced using surplus electrical power, production and storage offer another chemical means of storing energy. The greatest advantage would be gained if a hydrogen economy developed in which hydrogen replaced fossil fuel as a common combustible fuel.

The storage systems based on these processes are all viable ways of storing electricity and many of them are commercially available. Each has different characteristics such as response time and storage efficiency, which helps differentiate the technologies and define their applications.

Some of these systems can deliver power extremely rapidly. A capacitor can provide power in 5 ms, as a superconducting energy storage system can. Flywheels are very fast too, and batteries should respond in tens of milliseconds. A CAES plant probably takes 2–3 min to provide full power. Response times of pumped storage hydropower plants can vary between around 10 s and 15 min. This technology is generally suitable for peak power delivery but less suited to fast response grid support.

The length of time the energy must be stored will also affect the technology choice. For very long-term storage of days or weeks, a mechanical storage system is best and pumped storage hydropower is the most effective provided water loss is managed carefully. Batteries are also capable of holding their charge for extended periods. Energy loss in other systems will make them less practical for long-term storage. For daily cycling of energy, both pumped storage and CAES are suitable while batteries can be used to store energy for periods of hours. Capacitors, flywheels, and superconducting magnetic energy storage are generally suited to short-term energy storage, although flywheels can be used for more extended energy storage too.

Table 1.1 Round-Trip Efficiency of Energy Storage Technologies	
Energy Storage Technology	Round Trip Storage Efficiency (%)
Capacitors	90
Superconducting magnetic energy storage	90
Flow batteries	90
CAES	65
Flywheels	80
Pumped storage hydropower	75–80
Batteries	75–90

Another important consideration is the efficiency of the energy conversion process. An energy storage system utilizes two complementary processes, storing the electricity and then retrieving it. Each will involve some loss. The round-trip efficiency is the percentage of the electricity sent for storage, which actually reappears as electricity again. Typical practical figures for different types of system are shown in Table 1.1.

Electronic storage systems such as capacitors can be very efficient, as batteries can. However the efficiencies of both will fall with time due to energy leakage. Flow batteries, where the chemical reactants are separated, perform better in this respect and will maintain their round-trip efficiency better over time. Mechanical storage systems such as flywheels, CAES, and pumped storage hydropower are relatively less efficient. However the latter two, in particular, can store their energy for long periods if necessary without significant loss.

All these factors must be taken into consideration when considering the most suitable energy storage technology for a given application. For large-scale utility energy storage, there are three possible technologies to choose between, pumped storage hydropower, CAES, and—at the low end of the capacity range—large batteries. Batteries can also be used for small- to medium-sized distributed energy storage facilities,[2] along with flywheels and capacitor storage systems. Meanwhile fast-acting, small superconducting magnetic energy storage units are being used to aid grid stability. Superconducting facilities have been considered in the past for large-scale energy storage too but they appear to be prodigiously expensive based on the technology available today.

---

[2]Distributed storage facilities may be used by utilities to improve local grid stability or they may be used by consumers to make their own supplies more secure.

Table 1.2 Global Energy Storage Capacity, Broken Down by Type		
Storage Technology	**Number of Projects**	**Installed Capacity (MW)**
Electrochemical	985	3105
Pumped storage hydropower	352	183,387
Thermal storage	207	3692
Electro-mechanical storage	70	2585
Hydrogen storage	13	18
Liquid air energy storage	2	5
Lithium ion battery	1	24
Total	1630	193,266
Source: US Department of Energy.		

GLOBAL ENERGY STORAGE CAPACITY

The installed base of energy storage across the globe has been collated by the US Department of Energy (US DOE). Figures from its energy storage database are shown in Table 1.2. Based on the figures in this table, the total number of storage projects across the globe in 2017 was 1630 and their aggregate installed capacity is 193,266 MW. The largest part of this 183,387 MW (95%) is made up of pumped storage hydropower plants. (Note, however, that the International Hydropower Association puts the global pumped storage hydropower total at 150 GW.) Pumped storage hydropower plants are both the largest and the most widely spread of all storage technologies. In terms of numbers of projects, however, the largest number, 986, belongs to electrochemical energy storage—batteries[3] — but the total installed capacity of these is only 3105 MW or an average size of 3 MW. The aggregate includes a few large and a large number of small battery storage facilities.

The other technology with a large number of projects is thermal storage with 207 projects and an aggregate installed capacity of 3692 MW. While some of these are thermal storage facilities associated with solar thermal power plants, many are simply units that can store and then release heat energy for heating and hot water. This table show 70 electro-mechanical energy storage projects, primarily flywheel systems with a combined installed capacity of 2585 MW, 13 hydrogen storage facilities, and 2 liquid air energy storage projects.

---

[3]Table 1.2 shows 985 electrochemical storage projects and one lithium ion project.

Table 1.3 Top 15 Nations by Storage Capacity		
Country	Number of Projects	Installed Capacity (MW)
China	94	32,104
Japan	90	28,506
United States	494	24,123
Spain	66	8121
Germany	76	7567
Italy	52	7133
India	18	6013
France	21	5834
South Korea	62	4991
Austria	19	4680
Switzerland	23	4530
United Kingdom	29	3253
South Africa	9	3212
Ukraine	3	3173
Australia	39	2893
Source: US Department of Energy.		

Table 1.3 lists the major energy storage nations, ranked by the installed capacity in each country. Top of the list is China with 84 projects and an installed capacity of 32,104 MW followed by Japan with 90 projects (28,506 MW) and the United States with 494 projects (24,123 MW). In each of these countries, the largest part of the capacity will be based on pumped storage hydropower but the large number of projects in the United States suggests that a number of other technologies are gaining ground there too.

All the other countries in the list have less than 10,000 MW of installed capacity. Most of these are also nuclear nations that have built storage capacity to support their nuclear plants. Spain is perhaps an exception. Pumped storage capacity has been growing here to support the country's wind and solar capacity.

CHAPTER 2

Pumped Storage Hydropower

Pumped storage hydropower, the most important form of electrical energy storage in capacity terms, is an energy storage system based on the technology of hydropower. A conventional hydropower plant with a dam and reservoir involves an element of energy storage. The reservoir will store water during the rainy season, providing an energy source that can be used during dryer periods when the flow in the river is reduced. A pumped storage hydropower plant builds on this principle in order to be able to store and release electrical energy over a daily rather than a seasonal cycle.

In order to achieve this, the plant cannot rely entirely on the natural flow of water. Instead a typical pumped storage hydropower plant will have two reservoirs, one high and one low. When power is needed, water from the upper reservoir is released to the power house of the plant where it drives the turbines and provides electricity. Meanwhile during periods of low demand spare electricity from the grid is used to pump water from the lower reservoir back into the high reservoir so that is available again for generation. Energy storage plants of this type rely on cheap or surplus electricity being available, power that can be used to store water in the upper reservoir.

Mountainous regions offer some of the best potential for pumped storage because of the high heads of water that can be exploited. One of the first plants of this type was built in Switzerland, probably toward the end of the 19th century or the beginning of the 20th century. The use of pumped storage grew during slowly the early years of the 20th century but it was during the middle of the century that the major expansion took place, particularly in the United States, Europe, and Japan. Many of the pumped storage hydropower plants that were built during this period were originally designed to complement nuclear power generation. A conventional nuclear power plant operates most effectively if it generates power continuously, operating in what is

Power System Energy Storage Technologies. DOI: https://doi.org/10.1016/B978-0-12-812902-9.00002-X

known as a base-load power generation. However these plants are usually very large power plants with generating capacities in excess of 1000 MW and there may not be a need for their power during low demand periods such as the night-time. Pumped storage hydropower plants can use the surplus power from these large-generating stations to store energy overnight by pumping water into the upper reservoir. This energy can then be made available to meet peak demand the following day.

More recently there has been a growing need for electricity storage to help manage renewable electricity generation from intermittent technologies such as wind and solar power. These are a good source of electricity while the wind blows or the sun shines, but when wind or sun is not available, they can supply no power. The same principle that was applied to nuclear can be applied here too. So long as there is sufficient intermittent renewable capacity, surplus power available when the wind blows and the sun shines can be stored in a pumped storage hydropower plant. This energy can then be used to provide power when the renewable energy is not available. Spain and Portugal, in particular, have expanded their pumped storage hydropower capacity in recent years to balance wind and solar power on their national grids.

Like hydropower itself, the technology for a pumped storage hydropower station is simple and well understood. Turbine designs can be traced back to the 19th century and since then they have been refined into extremely efficient devices. Dam and reservoir construction dates back much earlier still: operating Roman dams can still be found in parts of Europe. The only unusual feature of the storage plant is that a pump is needed to move water into the upper reservoir in order to store energy. Early plants used separate pumps and turbines but most modern plants use reversible pump turbines.

While the technology presents no challenges, the civil works involved in the construction of a hydropower plant with two reservoirs, one high and one low, can be costly unless there are natural features that lend themselves to development. This is rare and more normally one of the reservoirs must be created. The construction work for such a plant makes the capital cost of the technology high. However once constructed, the lifetime of the plant should be measured in centuries rather than decades. Some pumped storage hydropower plants, provided they are built on a river, can also operate as conventional hydropower stations but many are designed specifically for energy storage alone.

Table 2.1 Global Installed Pumped Storage Hydropower Capacity by Region, 2016[1]	
Region	Installed Capacity (MW)
Africa	3376
East Asia and the Pacific	64,684
Europe	50,467
North and Central America	22,618
South America	1004
South and Central Asia	7541
Total	149,690
Source: From IHA.	

The total global installed pumped storage hydropower capacity at the end of 2016 was 149,690 MW based on figures from the International Hydropower Association (IHA),[2] as shown in Table 2.1. (As noted in Chapter 1, the US Department of Energy database records a higher global capacity.) The largest number of plants is in Asia, where Japan and China both have significant capacity, and in Europe and North America. Pumped storage plants also account for around 95% of the total active electricity storage capacity available on grids across the world according to the US Department of Energy. According to the Energy Storage Association, pumped storage plants in the United States can store around 2% of the nation's power output. In Europe the equivalent figure is 5% and in Japan storage plants can accommodate output from 10% of the country's power plants.

PUMPED STORAGE HYDROPOWER PLANT DESIGN

A typical pumped storage hydropower plant will have two reservoirs or basins that are connected via tunnels and shafts through which water can be passed from one to the other. Within the tunnel system there will be one or more hydro turbines and pumps (these will usually be the same unit) and there will be valves to control the flow of water from one reservoir to the other. A key feature of any hydropower plant including a pumped storage hydropower plant is the height of the column of water that is available to drive the turbine(s) in the plant. This is known as the head of water. For a given water flow, more energy can be extracted

---

[1]International Hydropower Association Hydropower status report 2017.
[2]International Hydropower Association Hydropower status report 2017.

from a high head than from a low head. This means that for pumped storage plants, high heads are preferred because they can be built with smaller turbines and pumps for a given generating capacity.

A schematic of a pumped storage hydropower plant is shown in Fig. 2.1. Within this overall design, there are several variations. The simplest type of pumped storage hydropower scheme is based around a conventional dam and reservoir power plant, one of the most common designs of hydropower plant in use today. These plants have a dam that is built across a river, impounding water behind it to create a reservoir. The reservoir stores water during periods of high river flow and this water can be released when flow levels are lower. A plant of this type can be converted into a pumped storage plant with the addition of a lower reservoir at the end of the outflow where water exits the power house. Under normal conditions, the plant will function simply as a conventional hydropower plant but with output regulated so that most power is supplied during periods of peak demand rather than operating as a base-load plant. In order to qualify as pumped storage such a plant must also be equipped with pumps that can pump water from the lower reservoir back into the main reservoir when water levels are low and surplus power is available. Such plants are often

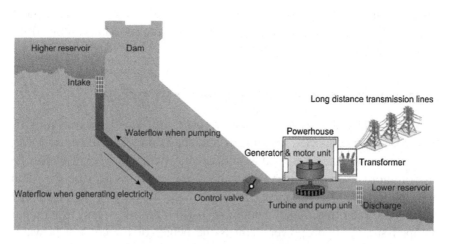

Figure 2.1 Schematic of a pumped storage hydropower plant. Source: University of Warwick.[3]

[3]Luo, X., Wang, J., Dooner, M. and Clarke, J. (2015). Overview of current development in electrical energy storage technologies and the application potential in power system operation. Applied Energy, 137, pp. 511–536. DOI: https://doi.org/10.1016/j.apenergy.2014.09.081.

called "on-stream integral pumped storage plants" or "pump-back pumped storage plants."

It may often prove convenient to make use of an existing dam and reservoir power plant in this way when planning a pumped storage hydropower plant because one of the reservoirs already exists. Many plants that are based on this design can operate both as pumped storage and generating plants. Frequently, however, both basins of a pumped storage hydropower plant need to be constructed. Where the topographical features exist, it may be possible to create the reservoirs from two existing waterways, provided they are separated vertically. The largest pumped storage hydropower plant in the world, Bath County in the United States with a generating capacity of 3000 MW, was created by building two dams, each damming an existing waterway to provide two new reservoirs. Water levels in both reservoirs vary during operation but the maximum head between the two is 400 m.

Sometimes it is more convenient to construct two man-made reservoirs that are not on any waterway (although some source of water will be needed to initially fill the reservoirs and to keep them replenished since there will be continuous losses). Old quarries have been used to provide one reservoir for such schemes. In other cases a reservoir has been built within a mountain. Since there is no natural source of flowing water sufficient for a conventional hydropower plant, plants of this type simply cycle water between and upper and a lower reservoir and can only function as storage plants. They are called closed-cycle pumped storage hydropower plants.

One of the most difficult tasks when considering construction of a pumped storage project is to find a location suitable for construction or exploitation of two reservoirs that are separated by a sufficient head to provide an efficient pumped storage plant. One option that has often been considered but rarely used is for the sea to provide the lower reservoir of a pumped storage scheme (see Fig. 2.2). For this to be effective there must be high cliffs adjacent to the shore and, crucially, either a lake or a site suitable for creating a lake at the top of the cliff. The largest plant of this type that exists today is a 30 MW scheme in Okinawa, Japan. Others have been proposed but none has yet been built.

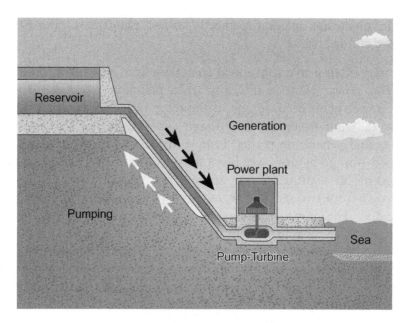

Figure 2.2 Schematic of a seawater pumped storage plant. Source: EnergyAustralia.[4]

Another alternative is to create a second reservoir underground. This is attractive because it has much less environmental impact than the creation of a new reservoir above ground but it is limited by the availability of a suitable underground reservoir. The most attractive sites are underground mine workings that have been abandoned. These exist in many parts of the world. However they have to be watertight to be of use, otherwise the water will leak away before it can be pumped back to the upper reservoir. Again, while such schemes have been proposed, none has been built.

Reservoir size is another factor that must be considered at the design stage, particularly if one or both are to be created. The storage capacity of the plant will depend on how much water the two reservoirs can hold. For a given set of turbines, the larger the reservoirs, the more hours of power can be stored. Some plants are designed with up to 20 h of storage capacity, others with as little as 4 h. When operated on a daily cycle, the round-trip efficiency of energy storage is 70%–80%.

Existing pumped storage plants have head heights of anything between 30 and 750 m. The minimum head height for a closed-cycle pumped storage plant is probably around 100 m but much higher

[4]https://www.energyaustralia.com.au/about-us/media/news/consortium-assessing-pumped-hydro-storage-plant-south-australia.

heads are advantageous since the equipment needed and the volume of water that has to be cycled is lower. Plants with the highest heads often require either separate pumps and turbines or multistage combined-pump-turbine units to be able to return water to the upper reservoir.

UNDERSEA PUMPED STORAGE

A novel proposal for pumped storage hydropower that avoids the need for reservoirs on land is to locate the storage under the sea. The concept is relatively simple. A strong container, probably constructed from concrete, is placed on the seabed. This container is fitted with a pump-turbine and then connected to the surface through a long pipe. The distance between the seabed where the container is located and the sea surface is the head of water available for the pumped storage facility.

Consider the seabed container as initially fully of seawater. During periods when there is surplus power, water is pumped out of the container, although the pipe to the surface then released into the sea. When power is required, valves allow seawater to flow down the pipe into the container, driving a turbine that generates electrical power. A system of this type requires a pump-turbine that is located in the container at the seabed.

This approach to pumped storage is attractive because it has minimal environmental impact. However it requires strong storage containers that can resist the pressure on the seabed. The deeper the seabed location, the smaller these need be for a given storage capacity. One of the most attractive applications of this type of technology is to link the seabed pumped storage units to offshore wind farms. The seabed units store surplus wind-generated electricity and then make it available when demand rises.

Seabed pumped storage appears to offer significant advantages over land-based systems but the technology has yet to be tested.

TURBINES AND MOTORS

A pumped storage hydropower plant must be able to both generate power from water running downhill and store energy by pumping water uphill. This requires both pumps and hydropower turbines. The most commonly used turbine for pumped storage hydropower is the Francis turbine. This is what is known as a reaction turbine. This type of turbine extracts energy from a head of water from the pressure that the head exerts onto its blades and, partly, from the kinetic energy of the flowing water as it strikes the blade. The Francis turbine was developed by American engineer James Bichens Francis around 1855. His design is extremely flexible and can be tailored to different head heights and flow rates. In operation the turbine must be completely immersed. The Francis turbine is also the basis for many water pump rotors. A modified version of the Francis turbine, called the Deriaz turbine, has blades that can be moved to change the angle of the blade meeting the water flow. Francis turbines are designed for specific head heights and do not operate with optimum efficiency if the head height varies significantly. The Deriaz turbine can maintain optimum efficiency under such conditions.

The second type of turbine that is used in some high head pumped storage hydropower plants is the Pelton turbine, patented by the American engineer Lester Allen Pelton in 1889. This turbine has bucket-shaded blades that are driven by the impulse from a fast flowing jet of water generated by the high pressure at the bottom of a head of water. This type of turbine, called an impulse turbine, is very efficient but cannot act as a pump.

The earliest pumped storage plants had separate pump-motor and turbine-generator units with each set mounted on their own shafts. The turbines could be either Pelton or Francis turbines depending upon the specific application, with the Pelton preferred for high head units. Some very high head pumped storage plants still use this configuration because the Pelton turbine, while the most efficient for exploiting a high head, cannot pump water back into the upper reservoir. For most, however, there was an element of redundancy with two motors and two turbines, one acting as a pump.[5]

By the middle of the 20th century, there had been some design consolidation. Since a motor is simply a generator running backward (and vice versa), the new configuration had a single shaft with a motor-generator at the top, a pump in the middle, and a turbine at the bottom. The motor-generator could then function with both pump and turbine. In many of these designs both the pump and the turbine were based on a Francis design. In spite of this, it was not until the middle

[5]The rotor in a pump is like a hydro turbine but operating in reverse.

of the century that single unit pump turbines began to be deployed in which the same rotor acted both as turbine and pump. These have now become standard for many pumped storage plants. However the configuration with a pump and a turbine on a single shaft is still used for some applications.

Pump turbines based on the Francis turbine can be used for a range of head heights. However they can be less effective for low head applications because in these the actual head height can change significantly as the upper reservoir is depleted and this will affect the generating efficiency of the turbine. A solution to this is to use a Deriaz pump turbine. As noted earlier, the Deriaz turbine is similar to a Francis turbine but had adjustable blades so that the blade angle can be altered as head height changes to maintain optimum efficiency.

Another advance that is being used in some modern plants is the variable speed turbine generator. Pumped storage turbines have traditionally been synchronized to the grid, so each time they start up they must be synchronized before they can begin to supply power. When synchronized, the generator of the system provides power at exactly the grid frequency. A variable speed turbine generator produces an output with a variable frequency depending upon its rotational speed. This type of generator uses power electronic circuitry to match the frequency of the power from the variable speed unit to that of the grid. Variable speed generators are more expensive than fixed frequency synchronized generators but they have advantages both in terms of ease of synchronization, and of efficiency because speed can be adapted to flow rate to maintain optimum efficiency.

Whatever type of turbine is used, pumped storage hydropower plants are capable of reacting extremely quickly to a demand for power. For example, the Dinorwig pumped storage hydropower plant in Wales has six 300 MW pump turbines that can synchronize and reach full power in around 75 s. What is more, if the turbines are first spun up in air and synchronized ready to deliver power, the full capacity of the plant—1800 MW—can be operational in 16 s.

In addition to being able to provide rapid response power to the grid, a pumped storage hydropower plant can help with grid stability in other ways. It can absorb surplus power rapidly as well as providing it, and can provide reactive load and voltage stabilization. It can also supply what is known as spinning reserve, providing the grid with inertia that helps resist sudden changes in conditions.

PLANT CAPACITY

Most pumped storage plants are large, often with generating capacities of several hundred or in the largest cases several thousands of megawatts. These large plants are expensive to build but can provide grid level storage on a very large scale. However it is not always easy or convenient to build a hydropower storage facility on this scale.

Pumped storage hydropower plants can be much smaller. Plants in the range 10– 50 MW of generating capacity are much cheaper and easier to build because sites are easier to identify. They can also be matched with single renewable generating facilities such as a large wind farm, making the power from the facility more easily dispatchable. However costs do not scale with size for hydropower; small hydropower plants are generally more costly per unit of generating capacity than large plants. This may make smaller plants uneconomical under many circumstances.

CHAPTER 3

Compressed Air Energy Storage

Compressed air energy storage (CAES) is an energy storage technology in which electrical power is used to compress air and the pressurized air is then stored in an air-tight chamber. This high pressure air is subsequently released through an air turbine to regenerate the electrical power. A CAES system is effectively a type of the gas turbine cycle but with the compression and expansion sections of the cycle decoupled one from the other.

As a means of both storing and distributing energy, compressed air systems have a long history. Networks for distributing compressed air were installed during the late 19th century in cities as various as Paris (France), Birmingham (the United Kingdom), Dresden (Germany), and Buenos Aries (Argentina) where they were used to supply energy to drive industrial motors and to support to a variety of commercial applications including the textile and printing industries. In these early applications the compressed air provided a medium for delivering mechanical power to users. Power delivery systems of this type became obsolete when electric power became widely available.

The use of compressed air storage as an adjunct to the power grid began with the construction of the Huntorf power plant that was built in Germany in 1978. However this plant only operated commercially for 10 years before being shut down. A second CAES plant was built by the Alabama Electric Cooperative in the United States and entered service in 1991. The facility has continued to provide storage services ever since. Details of these two facilities are shown in Table 3.1. Other plants have been planned, including one in Iowa called the Iowa Stored Energy Park with a proposed generating capacity of 270 MW. This project was eventually abandoned because the underground storage that was available at the site was not considered suitable.

Based as it is around the gas turbine cycle, a CAES plant is relatively simple to construct using conventional or off-the-shelf components. However, in spite of being championed by organizations such as the

Power System Energy Storage Technologies. DOI: https://doi.org/10.1016/B978-0-12-812902-9.00003-1

Table 3.1 Commercial CAES Plants

Plant	Generating Capacity	Commissioning Date
Huntorf, Germany	290	1987
McIntosh, Alabama, United States	110	1991

US Electric Power Research Institute, no further commercial project has ever been built although several others have been proposed and work even started on two. Even so, CAES remains of interest because it is the only other very large-scale energy storage system after pumped storage hydropower. Individual CAES plants are generally smaller than typical pumped storage plants but sites suitable for their construction are much more widespread than those for the hydro storage plants. The best large-scale storage caverns for CAES are underground geological features such as aquifers that can storage large volumes of gas. These are widely distributed across the globe and they could provide a large-scale energy storage network in most nations and regions.

THE COMPRESSED AIR ENERGY STORAGE PRINCIPLE

A CAES plant requires two principal components, a storage vessel in which compressed air can be stored without loss of pressure and a compressor/expander to charge the storage vessel and then extract the energy again. (The latter will usually be a compressor and a separate expander although in principle it could be a single unit.) In operation the plant is broadly analogous to the pumped storage hydropower plant. Surplus electricity is used to compress air with the compressor and the high pressure air is stored within the storage chamber. This stored energy can then be retrieved by allowing it to escape through the expander, an air turbine. The expanding air drives the air turbine that turns a generator to provide electrical power.

The compression and expansion functions of the CAES plant can be performed by two components of a gas turbine. The standard gas turbine comprises three major components, a compressor, a combustion chamber, and a turbine, with the compressor and turbine mounted on a single shaft. In operation air is drawn into the compressor and compressed, then delivered into the combustion chamber where it is mixed with fuel and the mixture ignited. This creates a hot, even higher pressure gas stream that then enters the turbine where energy is extracted to drive the shaft. In a modern gas turbine the turbine produces enough energy both to drive the compressor on the same shaft and to drive a generator that produces electrical power.

In a CAES plant the compressor and the turbine are separated and operate independently. CAES plants can also do without the combustion chamber. In this simple arrangement, shown in Fig. 3.1, air is drawn into the compressor that is driven by a motor running on surplus power from the grid. The compressed air from the output is then delivered into an underground cavern where it is stored. When power is required, air is withdrawn from the cavern and used to drive a turbine train, in this case a high pressure and a low pressure turbine on a single shaft with a generator. Any heat left in the exhaust air is captured in a recuperator and used to heat the air as it exits the cavern. This improves overall efficiency of the cycle.

In practice a CAES plant may prefer a slightly different configuration that is closer still to the gas turbine. A rotary gas turbine compressor is used to compress air that is then stored in the storage chamber. When power is required, compressed air is extracted again and fed into a combustion chamber where it is mixed with fuel and ignited, generating a higher pressure, higher temperature thermodynamic fluid that is then used to drive the turbine stage of the plant.

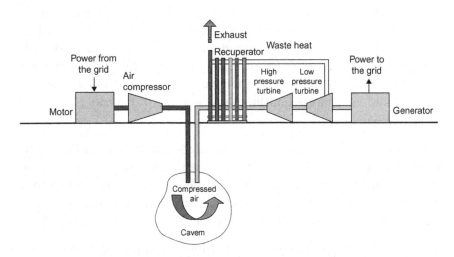

Figure 3.1 Schematic of a CAES plant.

Since a plant operated in this way requires natural gas or another fuel, it is not a straightforward energy storage system. However the economics of this mode of operation appears to be the most attractive because it can generate more electricity than was used to store the compressed air. Additional generation is between 25% and 60% depending upon the plant design. A further advantage is that since the turbine stage of the plant does not have to drive the compressor as it would in a conventional gas turbine, it can generate up to three times more power than it would when coupled to a compressor. Turbines for CAES plants can be relatively smaller than for a similar generating capacity gas turbine.

COMPRESSED AIR STORAGE

The most important part of a CAES plant is somewhere to store the compressed air. Small-scale CAES plants—with storage capacities of up to 100 MWh and outputs of up to 20 MW—can use above ground storage tanks built with steel pressure vessels but large, utility scale plants need underground caverns in which to store the air. The natural gas industry has used underground storage caverns for years to store gas; these same caverns can provide ideal storage facilities for a CAES plant. However the requirement for such a cavern limits the development of CAES to place where underground storage caverns are available.

A number of different types of underground cavern can be exploited. The most expensive is a man-made rock cavern excavated in hard rock or created by expanding existing underground mine workings. Such a site must be located in an impervious rock formation if it is to retain the compressed air without loss so the suitability of underground coal mines and limestone mines will depend on whether they are air-tight.

Salt caverns are another type of storage site, one that has been commonly used for gas storage. These are created within naturally occurring underground salt domes by drilling into the dome and pumping in water to dissolve and remove the salt to create an enclosure. Salt deposits suitable for such caverns occur in many parts of the world.

It is another type of geological structure, however, an underground porous rock formation that offers the cheapest underground storage facility. Structures of this type suitable for gas storage are found where a layer of porous rock is covered by an impervious rock barrier. Examples can be found in water-bearing aquifers, or in porous underground strata from which oil or gas has been extracted. Aquifers can be particularly attractive as storage media because the compressed air will displace water within the porous rock, setting up a constant pressure storage system. With rock and salt caverns, in contrast, the pressure of the air will vary as more is added or released. (This is referred to as a constant volume storage system.)

All three types of underground storage structure require sound rock formations to prevent the air from escaping. They also need to be sufficiently deep and strong to withstand the pressures imposed on them. It is important, particularly in porous rock storage systems, that there are no minerals present that can deplete the oxygen in the air by reacting with it. Otherwise the ability of the air to react with the fuel during combustion will be affected, reducing the power available during the generation phase of the storage-generation cycle. (This is only relevant for plants that use combustion to boost power output.)

Underground rock structures capable of storing compressed air are often widely available. For example, a survey in the United States found accessible sites of different types across 80% of the country.

A constant pressure storage system can also be set up using a form of underwater storage. In this case the storage vessel is an expandable container that is located a depth in the sea, or in a deep lake (see Fig. 3.2). When compressed air is pumped into the vessel, the hydrostatic pressure of the water will control the pressure of the gas and as more air is pumped into the vessel, so it will expand to contain it. The technology for the expandable vessel will be key to the effectiveness of this type of CAES storage.

Currently the cost of the storage vessel is considered to be the main factor limiting the development of CAES technology. If ready-made storage volumes are available, then it can be cost effective. If the storage vessel has to be created then costs may become overwhelming. The storage volume can vary depending upon the application. To be economically attractive, CAES plants are expected to have storage volumes equivalent to 500 and 2500 MWh of power and capacities of 50–300 MW.

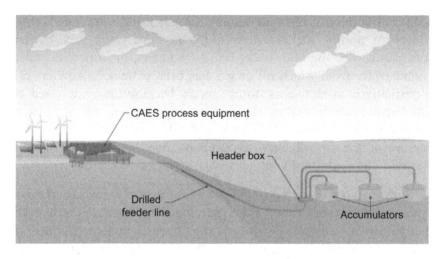

Figure 3.2 Underwater compressed air energy storage. Source: Hydrostor.[1]

TURBINE TECHNOLOGY AND COMPRESSED AIR ENERGY STORAGE CYCLES

A CAES plant generally exploits the standard gas turbine compressor and air turbine technology[2] but because the two units operate independently, they can be sized differently in order to match the requirements of the plant. The larger the compressor compared to the turbine, the less time it requires to charge the cavern with a given amount of energy. The Hundorf plant that was built in Germany required 4 h of compression to provide an hour of power generation, whereas the McIntosh plant in Alabama needs only 1.7 h of compression for an hour of generation.

As a consequence of compression and generation being separated, a CAES plant turbine can operate well at part load as well as full load. More complex operation is also possible. The Alabama plant, for example, uses two turbine stages with the exhaust from the last turbine used to heat air from the cavern before it enters the first turbine. The plant burns natural gas to increase the power output but fuel is not actually burnt in the compressed air until it enters a combustion chamber between the first and second chambers.

[1]https://hydrostor.ca/resources/Hydrostor-AECOM%202016-10.pdf. Main Hydrostor website: https://hydrostor.ca/.
[2]High pressure air storage will require a different sort of air turbine. This may be based on a steam turbine that can handle the higher pressures.

There are a range of more complex gas turbine cycles and CAES plants can utilize these too. For example, the compressor can be divided into two sections with air cooling between the stages to reduce its volume, a process called intercooling. Further cooling can be carried out at the exit of the compressor. The turbine stage can be adapted as well. One strategy is called recuperation. This involves capturing any remaining heat in the exhaust from the air turbine and using it to preheat the air from the storage chamber before it enters the turbine or turbines. The McIntosh plant, which has two turbines, uses another strategy called reheating. This involves burning fuel to heat the air between the turbines. As with the other strategies, this will increase efficiency but will cost more.

The air pressure of the compressed air that is stored in a CAES plant can be as high as 70 bar, much higher than the inlet pressure for typical gas turbines, which is around 11 bar. A conventional air turbine from a gas turbine cannot handle such pressures easily. The Huntorf CAES plant used an air turbine that was based on an intermediate pressure steam turbine since steam turbines are designed to operate at much higher pressures than gas turbines. In CAES plants that use two air turbines, a conventional gas turbine can act as the low pressure unit.

In principle a CAES plant could be of virtually any size and one proposed project would have had a generating capacity of 2400 MW. However in practice most schemes are likely to be smaller than this, in the tens or hundreds of megawatts range. Start-up time for the two plants that have operated was around 12 min but both could be brought into service in 5 min if necessary. Round trip efficiency without the use of additional fuel will be low for conventional CAES plants such as the two that have operated but refinements could improve this. For example, the Huntorf plant operated at an efficiency of 42% while the McIntosh plant has an efficiency of 54%.

ADIABATIC COMPRESSED AIR ENERGY STORAGE

One of the most important CAES cycle refinements is to build what is known as an adiabatic CAES plant. When a gas is compressed, its temperature rises as well as its pressure. Adiabatic compression and expansion involves retaining all the heat energy within the gas.

However in most CAES plants the heat energy produced during compression is lost, leading to a loss in efficiency.

An adiabatic CAES plant seeks to capture this heat during the compression stage and store it, then use it to reheat the gas when it is withdrawn from the storage chamber to produce electric power. This adiabatic cycle could theoretically be used to design a CAES plant that has no need for additional fossil fuel and that can achieve a round trip efficiency of 65%.

The key to adiabatic CAES is the ready capture and release of heat energy from the air. This is carried out using heat exchangers to extract heat from the hot compressed air before it is stored and then return it to the air extracted from the store. One experimental system of this type uses an oil heat exchanger in which cold oil passes into the device and emerges hot. This hot oil is then stored in a hot oil storage vessel. This hot oil is then passed through a second heat exchanger when cold compressed air is withdrawn from the compressed air store and transfers the heat energy back to the air. The round cycle efficiency of heat capture and delivery typically 70%–75%.

As will be clear, adiabatic CAES involves both compressed air and heat storage. This increases the complexity of the cycle significantly. However it does promise significant energy savings and if these can be realized then it may improve the economics of the technology.

ISOTHERMAL COMPRESSED AIR ENERGY STORAGE

An alternative to adiabatic CAES is isothermal CAES in which the air is compressed and expanded extremely slowly. If this is carried out carefully, then the heat that is lost to the atmosphere during compression is recovered during the expansion phase. In principle this can achieve 100% heat efficiency although in practice it will be lower. In systems of this type the temperature of the compressed air is always maintained within a few degrees of ambient temperature.

LIQUID AIR ENERGY STORAGE

Another technology that can be considered a form of CAES is liquid air energy storage. In this type of storage system the air is stored, not in a compressed gaseous state but in the liquefied state. This increases the energy storage density of the stored air by at least 10 times. In principle, for a plant of similar storage capacity, a liquid air energy storage system will be 10 times smaller than a conventional CAES system and 140 times smaller than a pumped storage hydropower reservoir.

A liquid air energy storage system uses off-peak power to compress, cool, and liquefy air. This air must then be stored in special cryogenic containers. Heat from compression may be captured and stored too if it is economic to do so. When power is required, liquefied air is released from the store and heated to regenerate the gaseous form. This provides a high pressure stream of gas that can be used directly to drive an air turbine.

One of the potential attractions of this type of system is that it can use waste heat from various sources to heat the liquefied air. This has the potential to increase the overall efficiency significantly. Waste heat might be provided by an industrial process or it might be the waste heat from the exhaust of a combined cycle power plant.

The liquefaction of air is a commercial process that is used in a variety of industries and the technology is well known. However it is expensive. The key to the economics of this technology is likely to be finding a more cost effective means of liquefying air than the methods currently available. Some companies have tested pilot-scale plants that exploit this technology but no commercial systems have been built.

CHAPTER 4

Large-Scale Batteries

Batteries are the most widely exploited means of storing electrical energy. Invented during the 19th century, the devices are used today for a whole range of portable applications from providing starter motor power in conventional vehicles to supply an electrical power source for mobile phones, tablet computers, and tiny electronic devices such as hearing aids. More recently batteries have also been used in a range of mostly small renewable energy applications and in addition some large cell stacks have been used for grid storage and stability applications.

Batteries are capable of reacting extremely quickly to changes in electricity demand and the type of cells that are used in conjunction with electricity generation can both deliver and absorb power rapidly. Today there is range of different types available, each with different properties. Many suffer from gradual degradation with prolonged use but the rate will vary from type to type. The advent of electrical devices such as portable computers and mobile phones that require reliable and long-lived electricity storage capabilities has pushed the development of battery technology in the past two decades and this has accelerated with the arrival of electric cars. Large-scale energy storage requires batteries with even more demanding specifications.

Batteries are electrochemical devices that convert the energy that is released during a chemical reaction into electrical energy. The chemical reactants are stored inside the battery case. The energy produced when they react would normally be released as heat if the reaction was permitted to proceed conventionally by mixing the reactants. In an electrochemical cell (another name for a battery) the reaction is controlled in such a way that most of this heat can be converted into electricity. In addition, the reaction can only take place when an electrical connection is made between the two electrodes of the cell. Without that the battery remains dormant.

Power System Energy Storage Technologies. DOI: https://doi.org/10.1016/B978-0-12-812902-9.00004-3

There are two distinct types of battery in common use, designated primary cells and secondary cells. A primary cell (or battery) can only be used once. After that it is spent. This type of cell is commonly used in a range of small portable devices but is of little use for large-scale applications. A secondary cell, in contrast, is reusable. It is capable of being recharged by applying a voltage across its terminals. This acts to reverse the internal chemical reaction that produced power and regenerate the reactants that provided that power in the first place. Secondary cells are the type that are used in the power and utility industries.

Secondary cells can be further divided into two types, standard secondary cells and flow batteries. Standard cells are the type that provide the power for portable phones and computers or drive the starter motor in a road vehicle. They are completely self-contained, and have no mechanical parts. Charging and discharging are carried out via the cell terminals and all of the reactants required are contained within the battery package. These standard secondary cells have two further variants, shallow discharge cells that are never fully discharged—such as those used for vehicle starter power—and deep discharge cells that can be completely exhausted and then recharged without damage.

A flow battery differs from a conventional secondary cell because the actual cell within which the chemical reactants react and generate electricity does not carry the reactants themselves. Instead these are stored in external reservoirs and pumped through the cell as required. This type of battery is more complex than a conventional secondary cell but it has the advantage that battery capacity is limited only by reservoir size and this can easily be increased for relatively little cost. On the other hand flow batteries usually need to be large to become economical because of the additional components required to manage the reactants. While a number of different flow batteries have been developed, they have had little commercial success.

THE BATTERY PRINCIPLE

A battery is a device that exploits the energy released in a chemical reaction to produce electricity. For this to be possible the reaction must be one that proceeds spontaneously under normal conditions. Such reactions generally release heat energy as they proceed but in a battery the reactants are only allowed to react in a specific way that requires some of this energy to emerge as electrical power. In principle 80% of more of the chemical energy that is released during the reaction can emerge as electrical power but it will vary with battery type, with rate of discharge and with a number of other factors. There are a range of chemical reactions that can be exploited to provide electrical power and the name of the battery is often derived from the reaction. For example, a lead-acid battery exploits a reaction between lead and sulfuric acid.

The chemical reaction used in every battery can notionally be divided into two half reactions and the battery will contain two electrodes, called the anode and the cathode; each electrode is associated with one of these half reactions. The reaction in a cell is often called a redox reaction because it involves reduction of one component and oxidation of the other. This is important because in the cell the reduction part of the reaction takes place at one electrode (the cathode) while the oxidation element takes place at the other (the anode). This is a key to cell operation.

The half reactions involve the creation of charged ions and the capture or release of electrons. Under normal circumstances where the reactants are intimately mixed, these processes occur simultaneously at the same location. However in a battery the two electrodes are separated by an electrolyte that will allow charged ions to pass from one electrode to the other but will not allow electrons to pass. These can only cross from one electrode to the second to complete the reaction through an external circuit. This is the electrical current that can be used to drive electrical and electronic equipment. It is this separation of processes by the use of a selective filter—in this case the electrolyte— that allows the cell to generate power. This is shown schematically in Fig. 4.1, which represents a rechargeable battery.

Consider as an example the lead-acid battery. In its charged state the anode of the cell is composed of lead, Pb, while the cathode is made of lead dioxide, PbO_2. The electrolyte into which these electrodes are immersed is made of sulfuric acid, H_2SO_4. When a connection is made between the electrodes, lead at the anode can react with the acid to form lead sulfate, $PbSO_4$, while hydrogen ions from the acid are released into the electrolyte solution and electrons are provided to the external circuit (the lead is oxidized). These electrons pass through

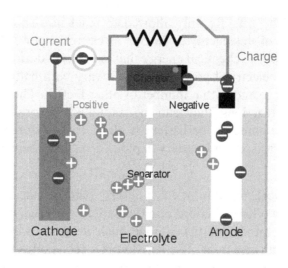

Figure 4.1 Schematic of a secondary battery. Source: Wikimedia.

the external circuit to the cathode where they are involved in a reaction between lead oxide and sulfuric acid to produce more lead sulfate and water (in this case the lead oxide is reduced). The formation of the water during the reaction depends on the hydrogen ions that are formed at the anode migrating through the electrolyte to the cathode. In both cases the lead sulfate that is formed remains attached to the electrode. Since the lead and lead compounds (Pb, PbO_2, $PbSO_4$) are isolated on the electrodes, the reaction can only take place with the combination of hydrogen ions passing through the electrolyte and electrons traveling through the external electrical circuit.

As already noted, batteries are divided into two types, primary cells and secondary cells. Primary cells contain reactants that will only react once to produce power. After that the cell is spent. This is fine for small cells that can be mass-produced. However it is neither practical nor economical for larger cells, or in many applications. It is also environmentally profligate because the primary cell is discarded once it is exhausted. A secondary cell can be recharged many times by applying a reverse polarity to the cell, reversing the chemical cell reaction and regenerating the original cell reactants. It is this type of cell that is of use for energy storage.

There are a number of battery types that can be used for grid and utility applications. The most frequently used large secondary cell is the lead-acid battery, similar to the type commonly used in vehicles.

Lead-acid batteries account for roughly half of all secondary cell sales, globally, and are widely available. Another type, nickel cadmium (Nicad or NiCd) batteries, is less common. NiCd batteries were frequently used for portable computers until they were superseded by alternatives which were better suited to the duty cycle of such devices. They have also been used for automotive applications, particularly under low-temperature conditions when they behave more reliably than lead-acid cells. Other cell types developed specifically for portable electronic devices include nickel-metal hydride cells and lithium ion cells. Both are potentially useful for utility storage and there are lithium ion cells being developed specifically for this application. A high-temperature battery, the sodium sulfur battery has also been popular for utility applications.

In addition to these conventional secondary cells, a variety of devices known as flow cells or flow batteries have also been tested for large-scale energy storage applications. These including zinc bromide, iron-chromium, vanadium redox, and polysulfide-bromide flow cells. None has so far found widespread commercial application but new types are being developed. They are considered attractive because they have potentially much longer lifetimes than conventional secondary batteries.

Traditional electrochemical storage systems boast a best case cycle conversion efficiency (electricity to chemical cell storage and back into electricity) of 90% but a more typical figure would be 70%. Most batteries also suffer from leakage of power over time. Left for too long, the cell discharges itself as a result of low-level electron conduction through the electrolyte. This means that battery systems will not normally be used for long-term energy storage. Flow cells do not suffer from this problem because the reactants are not stored together and this helps reduce long-term energy losses.

An additional problem with traditional secondary batteries is their tendency to age. After a certain number of cycles, the cell stops holding its charge effectively, or the amount of charge it can hold declines. Much development work has been aimed at extending the lifetime of electrochemical cells but this remains a problem. Again flow cells, because of their design, can avoid this problem.

Against this, batteries are able to supply their output extremely quickly, in under 5 ms for a conventional battery and less than 100 ms for a flow battery. Some are also capable of very high-power outputs and discharge rates.

LEAD-ACID BATTERIES

Lead-acid batteries were among the first secondary cells to be developed and were used for load leveling in very early power distribution systems. The cell is based on a reaction between lead, lead oxide, and sulfuric acid. Efficiencies of lead-acid batteries vary depending on factors such as the temperature and the duty cycle but are typically between 75% and 85% for DC to DC cycling. However cells discharge themselves over time so they cannot be used for very long-term power storage. If cycled carefully, cells for utility applications can have lifetimes of 15−30 years.

The cells have a water-based liquid electrolyte and operate at ambient temperature. Both high and low temperatures can reduce their performance. They are also relatively heavy and have a poor energy density although neither of these factors are a handicap for stationary applications. In addition, they are cheap and easily recycled.

Standard cells (often called flood cells) use a liquid electrolyte and must be maintained in an upright orientation to avoid electrolyte spillage. They release small amounts of hydrogen over time, so ventilation is important, and the level of the electrolyte must be topped up regularly too. An alternative design called the valve regulated lead-acid (VRLA) battery uses advanced design to allow the cell to place in different orientations without problem. Cells of this type use either a silica gel or a fiber-glass matt to contain the electrolyte between the electrodes. While this stabilizes the electrolyte, it does restrict the speed at which the cell reaction can take place. There is some electrolyte loss through hydrogen production but it is much smaller than in standard cell. However while electrolyte can be replaced in a standard cell, it cannot in a VRLA. These cells usually compensate by having an excess of electrolyte. VRLA cells are not usually suitable for applications which require high surge currents but they are used in energy storage where they are easier to maintain than flood cells.

Several very large energy storage facilities base on lead-acid batteries have been built. These include and 8.5 MW unit constructed in

West Berlin in 1986 while the city was still divided into East and West and a 20 MW unit built in Puerto Rico in 1994. While the former operated successfully for several years, cell degradation led to the latter closing after only 5 years. Lead-acid cells have been very popular for renewable applications such as small wind or solar installations where they are used to store intermittently generated power in order to make it continuously available.

NICKEL CADMIUM BATTERIES

The nickel cadmium battery is one of a family of nickel batteries that includes nickel-metal hydride, nickel iron, and nickel zinc batteries. There is also a nickel hydrogen battery in which one cell reactant is gaseous hydrogen. All have a nickel electrode coated with a reactive and spongy nickel hydroxide while the cell electrolyte is almost always potassium hydroxide. Cell reactions vary depending upon the second component. The NiCd battery was invented at the end of the 19th century but it was only in the 1960s that it began to be produced in large numbers.

The only nickel-based cell that has been exploited for utility applications is the nickel cadmium cell. Nickel cadmium batteries have higher energy densities and are lighter than lead-acid batteries. They also operate better at low temperatures. However they tend to be more expensive. This type of battery was used widely in portable computers and phones due to their higher energy density but it has now been superseded by lithium ion batteries.

NiCd batteries cool as they are recharged because reversing the cell reaction absorbs heat energy. This allows the cell to be charged very quickly, with full charge in 2 h or less, Efficiencies of nickel cadmium cells are typically around 70% although some have claimed up to 85%. However efficiency appears to vary with charging rate, the faster they are charged, the better the efficiency. Early cells suffered from a memory effect in which the cell would "remember" how far it was discharged during the previous cycle and only discharge to this level, reducing the overall charge available. Modern cells have reduced the memory effect, but have not eliminated it. Cells tolerate deep discharge well, which is good for energy storage applications. Lifetimes tend to be rated at around 10–15 years though some have lasted longer. These cells self-discharge themselves more rapidly than lead-acid cells and can lose 5% of their charge in a month. There can also be a problem with disposal since cadmium is highly toxic.

The largest nickel cadmium battery ever built is a 40 MW unit in Alaska, which was completed in 2003. It occupies building the size of a football field and comprises 13,760 individual cells.

LITHIUM BATTERIES

Lithium batteries including both lithium hydride and lithium ion batteries have become popular for consumer electronic devices because of their low weight, high energy density, and relatively long lifetimes. Lithium is extremely reactive and can burst into flames if exposed to water but modern lithium cells used lithium bound chemically so that it cannot react easily. As with nickel, there are a number of lithium cell variants but the most popular today is the lithium ion cell. These are designed so that there is no free lithium present at any stage during the charging or discharging cycle. The lithium cell was first proposed during the 1970s but did not become popular until late in the first decade of the 21st century.

The lithium ion cell involves a reaction between metallic lithium and a metal oxide. This is usually cobalt oxide (CoO_2) in small devices but the CoO_2 design can be dangerous if damaged and large cells may use a different metal oxide. The metallic lithium is usually contained with interstices in a carbon or graphite electrode from which it can migrate easily. During operation lithium ions are formed and these migrate through the electrolyte to the other electrode where they enter the metal oxide structure and form lithium cobalt oxide ($LiCoO_2$). The electrolyte in the cell is usually a lithium salt in an organic solvent which provides a migratory path for the lithium ions. A simple schematic of a lithium ion cell is shown in Fig. 4.2.

Lithium ion cells have a low self-discharge rate, a high energy density, and a very small memory effect. This has made them popular for portable applications such as phones and computers. There are also increasingly used in place of lead-acid batteries in luxury products such as golf carts. More recently there has been growing interest in their use for energy storage.

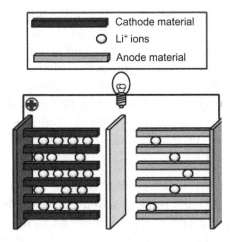

Figure 4.2 Schematic of a lithium ion cell. Source: Wikimedia.

The use of lithium batteries in grid and utility applications is beginning to grow with units being tested in a number of locations. An early large pilot battery storage installation rated at 2 MW was commissioned on the Orkney Islands which are located off the coast of north western Scotland, in 2013. This was topped in 2017 when the US utility San Diego Gas and Electric opened a 30 MW battery storage facility based on lithium ion batteries with 120 MWh of storage capacity. A 20 MW facility is also being planned by the utility Southern California Edison. The future development of lithium batteries may benefit from interest by automotive manufacturers in their use in hybrid and electric vehicles.

SODIUM SULFUR BATTERIES

The sodium sulfur battery is a high-temperature battery. It operates at 300°C and utilizes a solid electrolyte, making it unique among the common secondary cells. One electrode is molten sodium and the other is molten sulfur and it is the reaction between these two that is the basis for the cell operation. Although the reactants, and particularly sodium, can behave explosively, modern cells are generally reliable. However a fire was reported in 2012 at a sodium sulfur battery installation in Japan.

In order to create a workable cell from these elements, the sodium and sulfur must be separated from each other by an impermeable electrolyte. This is a solid, beta alumina, which will prevent the molten sodium from passing through while allowing positively charged sodium ions to travel to the sulfur electrode. The molten sulfur, meanwhile, is contained in a carbon sponge. In practical cells the molten sodium is contained in the center of the cell, surrounded by the solid electrolyte, while the sulfur containing carbon sponge sits around this. The whole is contained in a steel case that is coated with chromium or molybdenum to protect it from corrosion. This case acts as the cathode of the cell. A cross-section of a sodium sulfur battery is shown in Fig. 4.3.

Early work on the sodium sulfur battery took place at the Ford Motor Co in the 1960s but modern sodium sulfur technology was developed in Japan by the Tokyo Electric Power Co, in collaboration with NGK insulators and it is these two companies that have commercialized the technology. Typical units have a rated power output of 50 kW and 400 kWh. Lifetime is claimed to be 15 year or 4500 cycles and the efficiency is around 85%. Sodium sulfur batteries have one of the fastest response times, with a startup speed of 1 ms. The sodium sulfur battery has a high energy density and long cycle life. There are programmes underway to develop lower temperature sodium sulfur batteries.

This type of cell has been used for energy storage in renewable applications. The largest installation to date is a 34 MW, 245 MWh

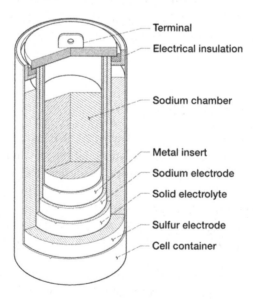

Figure 4.3 Cross-section of a sodium sulfur battery. Source: NASA.

facility in Japan that is used for grid support to provide wind energy stabilization. There are some large units in the United States too, including a 4 MW battery storage system at Presidio, Texas, which can store up to 4 MW of power for 8 h (32 MWh) in the event of grid failure. The fast response time and good cycle life make them attractive for grid support and energy arbitrage too. However the technology is complex and small-scale batteries based on sodium sulfur are less economical.

FLOW BATTERIES

A flowing-electrolyte battery, or flow battery, is a cross between a conventional battery and a fuel cell. It has electrodes like a conventional battery where the electrochemical reaction responsible for electricity generation or storage takes place, and an electrolyte (an ion exchange membrane). However the chemical reactants responsible for the electrochemical reaction and the product of that reaction are stored in tanks separate from the cell and pumped to and from the electrodes as required, much like a fuel cell. The layout of a typical flow battery is shown in Fig. 4.4.

The principles of operation of the flow battery are identical to those of a conventional electrochemical cell with a reaction mediated by ions that pass from one electrode to the other through a membrane that only

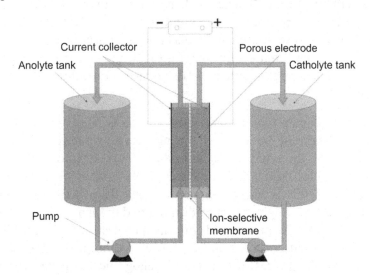

Figure 4.4 Schematic of a flow battery. Source: Wikimedia.

lets particular ionic species pass. Electrons must travel through an external circuit to complete the reaction. The advantages of the flow battery over a conventional secondary cell are longevity and capacity. Reactants can be renewed easily without dismantling the complete device. However the devices are more complex, with mechanical components such as pumps. This increases the basic cost and means that flow batteries generally need to be large to be economical. Response times for flow batteries are longer than for conventional secondary batteries but they should be able to supply full power within 100 ms. The cells can have high-power and current densities but their energy densities are lower. However they are considered well suited to electrical energy storage. Their use in electric vehicles has also been examined.

A range of chemical reactions can be utilized in a flow battery. At least 15–20 have been tried. The most prominent so far have been the zinc-bromide flow battery, the polysulfide-bromide battery, and the vanadium redox battery. A number of newer designs are also in the research stage including versions based on organic compounds. However flow batteries have not been tested commercially so their overall performance has yet to be established.

ELECTRIC VEHICLES

The advent of electric vehicles with battery storage could potentially revolutionize energy storage. The number of battery vehicles, globally, is still small but environmental concerns are pushing manufacturers to develop vehicles that do not rely on fossil fuel and the battery vehicle is one of the prime contenders to replace the traditional petrol or diesel vehicle.

While the capacity of an individual vehicle is relatively small, once the vehicles become widespread the aggregate capacity will become enormous. This becomes significant when you take into account that all these vehicles have to be attached to the grid in order to charge their cells and this is like to happen at least daily. Much recharging will take place overnight but even during the day there will be a significant battery capacity on the grid. In principle electric cars with fully or partly charged batteries that are connected to the grid could be used for grid support services such as load leveling and frequency and voltage stability.

A lot of elements would need to be in place to make this feasible including charging hardware that can be used by the grid controller and tariffs that encourage users to hook up their battery vehicles when not in use. The population of such cars it still too small to establish whether this is truly workable but the concept certainly has much to recommend it.

CHAPTER 5

Superconducting Magnetic Energy Storage

Superconducting magnetic energy storage (SMES) is a technologically advanced method of storing energy in a magnetic field, which is formed when a current flows around a coil. In order for this to operate efficiently as an energy storage system, the coil must be made of a superconductor that has no electrical resistance so that there are no resistive energy losses as the current circulates. The superconducting materials needed to create the energy storage coil are expensive and the best of them available today must be cooled cryogenically to close to absolute zero temperature before they become superconducting. Higher temperature superconductors are available too but they tend to be less effective.

Storage of energy based on superconducting rings offers perhaps the highest round trip efficiency available of all energy storage systems. However the cost of maintaining the extremely low temperature of the coil increases the operating costs over time and reduces overall efficiency. Even so, SMES has been proposed for load leveling services similar to those offered by pumped storage hydropower plants. SMES is also one of the fastest acting and highest current density storage systems available.

Practical SMES energy storage systems for grid applications were developed at the end of the 1980s. These are relatively small and have been used for grid support and power conditioning. Most can supply 1 MW or less of power. Larger systems have been tested and there have been proposals for massive SMES storage systems, capable of providing an output of 1000 MW or more. However the physical structure needed to contain the enormous magnetic forces generated by such a coil would be massive and such large systems have not so far proved to be economical enough to consider construction. To date, it is the smaller systems that have found commercial success, normally in grid support functions or for power conditioning.

Power System Energy Storage Technologies. DOI: https://doi.org/10.1016/B978-0-12-812902-9.00005-5

THE SUPERCONDUCTING ENERGY STORAGE PRINCIPLE

Superconductivity offers, in principle, the ideal way of storing electric power and one of the few ways in which electromagnetic energy is stored directly. The storage system comprises a coil of superconducting material that is kept extremely cold. Off peak electricity is converted to direct current (DC) and fed into the storage ring using a converter system and there it stays, ready to be retrieved as required. The energy is actually stored as a magnetic field that keeps the current circulating; in effect the two are mutually self-supporting and with no resistive energy loss, the current will continue to circulate virtually indefinitely. Provided the system is kept below a certain temperature, energy stored in the ring will remain there without loss.

The key to the SMES device is a class of materials called super-conductors. Superconductors undergo a fundamental change in their physical properties below a certain temperature called the transition temperature, which is a characteristic of each material. When a material is cooled below its transition temperature, it becomes superconducting. In this state it has zero electrical resistance. This means that it will conduct a current with zero energy loss.[1]

Superconductivity was discovered in the early part of the 20th century. It is a quantum effect and cannot be understood using classical physical concepts of resistance. A normal conducting material such as copper will still have a small electrical resistance when it is cooled close to absolute zero due to impurities in the material. However in a superconductor, once the temperature drops below the critical temperature, its electrical resistance falls abruptly to zero and a new quantum state is set up within the superconducting material. The best metallic superconducting materials available today must be cooled to a temperature close to absolute zero before they become superconducting. The most widely used and cheapest superconducting material is a niobium-titanium alloy that has a critical temperature of $10°K$ or $-263°C$. Liquid helium is usually used as the coolant to cool a coil made of the alloy below this temperature. Niobium-titanium can support relatively high magnetic fields. Other more expensive materials are also available if an application demands very high magnetic fields.

---

[1] The device used to charge and withdraw energy from the superconducting coil does have a slight resistance and this leads to small energy losses over time.

A superconducting current can only be maintained in a coil by keeping the coil below its transition temperature and this requires extreme refrigeration. The cooling system accounts for one of the main running costs of such a system. High-performance insulating materials can help keep heat transfer rates low.

In recent years scientists have discovered a new range of ceramic materials that become superconducting at relatively high temperatures, temperatures accessible by cooling with liquid nitrogen. (Liquid nitrogen boils at 98°K, −175°C.) Most of these materials have proved to be rather brittle ceramics that are difficult to work but techniques are being found to exploit them. This is helping make superconductivity more economically attractive for a range of utility applications including transmission lines and storage.

Fig. 5.1 shows a diagram of a typical small commercial SMES system that might be used for grid stability applications. In this type of system the small SMES storage ring is built into a container for ease of transportation and installation. The storage ring is connected to the electricity distribution system at the site where it is needed. Power from the grid is used to drive the cooling system that maintains the actual storage ring below its critical temperature. The SMES ring is then charged from the grid. In this case the ring will be used for grid stability support or power conditioning. Sensors will detect the condition of the grid and if there are frequency, voltage, or phase fluctuations, power can be drawn from the ring to correct the instability.

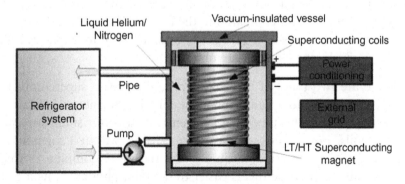

Figure 5.1 Schematic of distributes SMES system. Source: University of Warwick.[2]

[2]Luo, X., Wang, J., Dooner, M. and Clarke, J. (2015). Overview of current development in electrical energy storage technologies and the application potential in power system operation. Applied Energy, 137, pp. 511–536. DOI: https://doi.org/10.1016/j.apenergy.2014.09.081.

Superconductors are charged with DC to create the magnetic field in which the energy is stored. There is no energy loss while the energy is in the coil but there are energy losses that occur during the conversion of grid alternating current (AC) to DC and then from DC back to AC. The round trip efficiency is typically 90% for daily cycling but will be lower for long-term storage because of the energy required to maintain the coil at below its transition temperature. There are also small continuous losses within the coil at the point where power is fed in and out. Start-up time for a SMES system is typically around 5 ms.

The advantages of SMES over other energy storage systems are primarily related to the speed of operation. A 50 Hz grid cycle takes 20 ms so that an SMES storage facility can intervene within the space of one cycle if it is required to do so. In addition, an SMES coil is capable of discharging all the energy it contains in a short space of time. This means that it can provide a high-current density, a facility that is not always available from other storage techniques.

One of the key questions regarding SMES is whether it is better to use high-temperature superconducting materials that require only liquid nitrogen to cool them, or low-temperature superconductors that need liquid helium or liquid hydrogen. While the high-temperature materials are ostensibly cheaper, their physical and electrical properties mean that in fact a lot more of this type of material is needed to provide a ring with the same storage capability as a ring using a low-temperature superconductor. In all but very small systems, the actual cost of the cooling system is small compared to the cost of the superconductor and cooling to close to absolute zero is not necessarily more expensive than cooling to 77°K, the temperature of liquid nitrogen.

There are two types of coil that can be used for a superconducting storage coil, a solenoid and a torroidal coil. The former, a helical coil, is the simplest and this type of coil is generally used for small superconducting coils. For larger coils the internal forces within a solenoid coil become too large to manage simply and the more complex torroidal coil is preferred because this is easier to contain, physically, as a result of the external magnetic forces being smaller.

One additional advantage of SMES over other types of energy storage is that an SMES coil has no moving parts so that it should have a very long lifetime. The forces within the ring resulting both from contraction as it cools to its operating temperature and magnetic forces when a current is circulating could cause fatigue fractures but provided the coil is designed to minimize or avoid these, then the lifetime could be similar to that of a pumped storage hydropower plant. The cooling system might need renewing over time, but the most expensive part, the coil should continue to be usable.

APPLICATIONS OF SMES

When SMES devices were first proposed, they were conceived as massive energy storage rings of up to 1000 MW or more, similar in capacity to pumped storage hydropower plants. One ambitious project in North America from the last century would have had a storage capacity of 2400 MW. This would have required a storage ring, buried underground, tens or perhaps hundreds of kilometers in diameter. The project was eventually deemed uneconomical due to the enormous construction cost and the cost of superconducting wire at the time. However the idea of a massive superconducting energy storage system has not gone away and the US Department of Energy is funding research into large-scale SMES storage.

Current technology allows small commercial SMES storage units with capacities of between 100 kW and 100 MW to be constructed. The largest built to date can deliver 10 MW. The storage capacity of these commercial devices is between 10 and 30 kWh, relatively low for utility storage but useful for very fast grid support functions. These are commonly referred to as micro-SMES systems.

One of the earliest SMES devices to be used commercially was commissioned by the US Bonneville Power Administration in the 1980s. This unit had a storage capacity of 30 MJ and a power rating of 10 MW. The device could release 10 MJ of energy in one-third of a second to damp power swings on the Pacific Intertie. Today a typical commercial unit has a storage capacity of 3 MJ (0.83 kWh) and can deliver 3 MW of power for 1 s.

Micro-SMES is being used in many parts of the world for power quality control where speed of operation is the main attraction. The efficiency of these small units is lower than could be achieved in a large coil but in this application efficiency is not of primary concern. These units can be fitted into a container, as shown in Fig. 5.1 for ease of deployment. They are used at substations and in industrial operations where high-quality power is required.

The next stage in the development of SMES is likely to be the use of larger superconducting coils for distributed generation system stability support. While micro-SMES systems can typically provide between 1 and 3 MW, distributed generation systems might be expected to be able to have outputs of 10 times this power. One ambition is to develop SMES systems of this scale that can compete with batteries for energy storage. However there are no commercial systems available in this power range.

The ultimate goal may be to develop even larger SMES systems for power and load leveling. Energy storage is required on grids across the world to help stabilize renewable input. Large SMES units with their ability to respond quickly would be ideal for this application. However the costs still appear prohibitive. One advantage that a large SMES system has over pumped storage hydropower, the main large-scale energy storage technology in use today is ease of siting. While pumped storage hydropower requires a suitable site for the construction of two reservoirs, a large SMES ring could be built at a range of sites, particularly where the land has no other use, such as in desert regions. Such regions are also potential sites for large solar plants and the combination of solar power, and SMES could provide a valuable source of dispatchable renewable power, provided costs can be lowered.

The future of large SMES may depend on the development—or otherwise—of cheaper superconducting materials, and particularly higher temperature superconductors with better properties. The ultimate goal would be to discover materials that are superconducting at ambient temperatures. While a number of tantalizing hints have appeared in scientific journals, that prospect still remains extremely distant.

CHAPTER 6

Flywheels

A flywheel is a simple mechanical energy storage device comprising a large wheel on an axle fitted with frictionless bearings. The flywheel stores kinetic energy as a result of its rotation. The faster it rotates, the more energy it stores. Provided there is a means both to store and then to extract this energy again, the system can be used for a variety of applications.

Traditional flywheel-based systems have been in use as mechanical energy storage devices for thousands of years. Millstones use stored energy to maintain their smooth operation while potters wheels and hand looms have all used flywheels to both store energy and smooth out the pulsed of power in energy delivery by hand or foot. Industrial-scale systems such as 18th century looms used them too. The flywheels were often massive and these systems were frequently powered by a water mill.

Another development in the 18th century, the steam engine, adopted the flywheel as a means of smoothing the pulses of energy from the steam piston. This has been carried into the 20th and 21st centuries: simple flywheel energy storage devices are fitted to all piston engines to maintain smooth engine motion. The engine flywheel is attached physically to the engine camshaft and as the pistons cause the camshaft to rotate they feed energy into the flywheel. The rotational energy stored in the flywheel helps drive the pistons between each stroke and smooths the otherwise intermittent power delivery. This energy is necessary to drive the pistons during the part of each cycle where they are not delivering power.

The application of flywheels to electrical energy storage began to be explored during the 1960s. Success has been erratic but at the beginning of the 21st century, these devices have become accepted as a means of providing short-term, ride-through power to electrical supply

Power System Energy Storage Technologies. DOI: https://doi.org/10.1016/B978-0-12-812902-9.00006-7

systems. Most flywheel storage systems store a very small amount of energy but this energy can be delivered very quickly. They are normally used supply transient grid support from the moment an interruption occurs until a secondary supply such as an engine can cut in. However some flywheels systems have been designed to supply their rated power for several minutes, or even for hours.

For electricity storage applications, energy will normally be fed into the flywheel using a reversible motor generator. Energy losses must be kept to a minimum so special frictionless bearings are required and the rotating device is usually housed in a vacuum chamber to reduce friction losses from air. The energy stored in a flywheel depends on the rotational speed and higher speeds are favored since high-speed devices can be smaller for a given energy storage capacity.

THE FLYWHEEL PRINCIPLE

A flywheel uses a rotating mass to store energy which is held in the kinetic energy of rotation of the rotor. The amount of stored energy is proportional to the moment of inertia of the rotor about its axis (a property directly related to its mass) and to the square of its rotational speed. Increasing either mass or rotational speed will increase storage capacity but higher speeds offer the more efficient way of raising capacity. However high speeds can make severe demands on the materials used in fly wheel construction.

Conventional flywheels such as those used on the crankshafts of piston engines are fabricated from heavy metal disks made of iron or steel. These disks are only capable of rotating at relatively low speeds, up to around 10,000 rpm. For power applications, new lighter composite materials have been developed, capable of rotating at 10,000–100,000 rpm without fracturing under the immense centrifugal force they experience. These special rotors are made from carbon fiber or glass fiber composite materials which while less dense that the metals they replace—and therefore having less mass—can store more energy at the very high rotational speeds at which they operate.

The rotor of a modern energy storage flywheel is usually a cylinder rather than a disc as this concentrates the mass of the rotor as far from the rotational center of the flywheel as possible. The moment of inertia is proportional to the square of the distance from the axis and so this maximizes rotational mass and hence energy storage. Fig. 6.1 shows a schematic of an energy storage flywheel with a cylindrical rotor.

One of the key sources of energy loss in a flywheel system is frictional loss associated with the rotor bearings. These must support the rotor while it turns and are key to its operation. Mechanical bearings are the simplest and best understood form of bearing but these also have the highest frictional levels and greatest wear rates. Such bearings may be used for low rotational speed flywheels but the high-speed devices require another type of bearing. However all flywheels are fitted with mechanical bearings too, for start-up and shut down.

The standard type of bearing used in high-performance flywheel systems is a magnetic bearing that allows the rotor to be supported during rotation without any physical contact with the support. The magnetic field is provided by an electromagnet and the bearings can either be passive, in which case they simply support the load of the rotor, or active magnetic bearings that can provide more intelligent control of the rotor as it spins. A particularly high-performance type of magnetic bearing relies on a superconducting magnet to provide the magnetic force. This can be extremely powerful but requires energy to maintain the magnetic coil at a very low temperature.

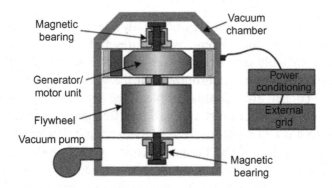

Figure 6.1 Schematic of an energy storage flywheel system. Source: University of Warwick.[1]

[1]Luo, X., Wang, J., Dooner, M. and Clarke, J. (2015). Overview of current development in electrical energy storage technologies and the application potential in power system operation. Applied Energy, 137, pp. 511–536. DOI: https://doi.org/10.1016/j.apenergy.2014.09.081.

The other source of friction in a flywheel system is air. As a fly-wheel rotor turns, the air layers close it start to move too, setting up a velocity gradient between air layers, creating drag and in some cases turbulence that will result in energy loss. As with bearing friction, the losses rise with the speed of rotation. To avoid this, modern high-performance flywheels are housed inside a vacuum chamber so that frictional air drag and loss is eliminated. As a consequence of the rota-tional speed, the whole device must be enclosed within an exceedingly strong container that will prevent the pieces of the flywheel scattering like shrapnel in the event of a catastrophic failure. Some lower speed wheels may still be operated in air.

The other key element of a flywheel storage system is a means of both charging the device with energy and then extracting it again. Energy is stored in the flywheel using an electric motor attached to the shaft upon which the flywheel is mounted. In a modern cylindrical flywheel this will often be inside the cylinder. Energy can be added and the rotational speed of the flywheel is increased until the device reaches its maximum speed.

Extracting this energy once more requires a generator. Like the motor this will be attached to the shaft of the device. It is possible to have a separate motor and generator but most flywheels will use a single device that can operate both as a motor and a generator to save cost and to simplify device design.

When energy is extracted from the flywheel, the rotational speed will fall as the amount of energy it stores falls. This variable speed results in a variable frequency power output from the generator that makes it impossible to synchronize with the grid using a conventional generator. To get around this problem the flywheel will use power electronic devices to convert the variable frequency output from the generator to a direct current (DC), which is then converted back to an alternating current (AC) at the grid frequency. AC−DC−AC conver-ters of this type allow energy to be both stored and extracted with ease. According to the US Department of Energy, energy can be stored and extracted using systems of this type with a round trip efficiency of 75%.[2] Some manufacturers claim that an efficiency of 90% is possible.

[2]Federal Technology Alert—Flywheel Energy Storage, US Department of Energy, Energy Efficiency and Renewable Energy, 2003.

Heat management is important for flywheel design because even when frictional losses are reduced by using magnetic bearings and operating in a vacuum, there is still some friction and this emerges as heat. Most of this heat energy will be in the form of radiant heat and flywheel design must ensure that this is managed so that overheating does not occur. Most require cooling systems.

The other operational issue with all flywheels is safety in the case of catastrophic failure. Energy storage flywheels even those operating at relatively low speeds below 10,000 rpm are subject to massive centrifugal stresses. These become much higher as the speed approaches 100,000 rpm and flywheels can fail, with the flywheel rotor simply breaking apart. If this happens, the pieces will fly out like shrapnel from a bomb. To avoid any danger as a result of failure, flywheel energy storage devices must be housed in a protective housing of steel. The latter can also double as the vacuum chamber. Catastrophic failure, although rare, is more commonly found with composite flywheels operating and very high rotational speeds. Steel flywheels that generally operate at lower speeds usually fail more slowly and faults can be detected before the device fails.

FLYWHEEL PERFORMANCE CHARACTERISTICS

In order for a flywheel to operate effectively as an electrical energy storage system, it must be kept rotating at its full operating speed. This requires continuous energy feed to compensate for friction losses, which will tend to slow the flywheel over time. The level of frictional losses will vary with design but are typically between 0.1% and 1% of the rated power. The energy required to maintain the device fully charged will mount over time and so the level of loss that can be tolerated will depend on the use to which the device is being put. Flywheels that are intended for ride-through grid support in times of a grid outage—bridging the short period between the failure and the point at which a backup system such as a diesel generator can come into operation—are generally conservatively rated and operate at relatively lower rotational speeds because security is more important than overall efficiency. With these devices, higher energy loss can be tolerated. On the other hand systems that are designed specifically for energy storage will operate at the highest speeds and be designed for the lowest overall losses to achieve the highest efficiency possible.

The energy stored in a flywheel is a function of the mass of its rotor. The larger the rotor, the more it can hold. However the power it can deliver will depend on the size of the generator used for energy extraction. Many flywheels are designed for backup power or grid support functions during which they are expected to deliver high power for a short period of perhaps a few seconds at most. Such units can have small rotors but relatively large power extraction systems. Other flywheels have been designed for long-term energy delivery. These will have rotors that are capable of storing a large amount of energy compared to the size of the energy extraction system.

The energy density of a flywheel is relatively high compared to battery systems with which they often compete. A flywheel system for backup services might achieve an energy density of 130 Wh/kg, similar to a sodium sulfur battery and much higher than a lead-acid battery. This, in turn, leads to a smaller footprint for the flywheel device. In addition a flywheel can be cycled more frequently than most battery systems. Modern flywheels claim to be capable of up to 175,000 full-depth discharge cycles, much higher than any battery-based system can achieve.

Flywheels can usually respond quickly when required, though not as swiftly as some battery systems. When used as grid backup systems, a flywheel should be capable of reaching full power within half a cycle, 25 ms, and some are quoted with response times of 5 ms. Such units will probably be able to supply their full output for between 5 and 15 s.

Commercial flywheels are available with power ratings of between 2 kW and 2 MW and with storage capacities of between 1 and 100 kWh. For larger power ratings, multiple flywheels are usually installed in parallel. One large commercial system is a unit with ten 100 kW flywheels used by the New York Transit System to support its electric traction power network. This system can supply 1 MW of power for 6 s. The largest flywheel energy storage device so far constructed is a device used in Japan for fusion research. This system can supply 340 MW for 30 s.

FLYWHEEL APPLICATIONS

There are a number of important applications for flywheels today. One of the most significant is for energy recovery in motor vehicles. When a vehicle brakes, the braking is carried out—partly at least—using a flywheel that absorbs the kinetic energy of motion and converts it into rotary motion. The energy can then be recovered using an electric generator that will in turn drive a motor that provides additional power to the vehicle. This type of application is similar in concept to grid energy storage.

Since the turn of the 21st century, flywheel energy storage systems have been used as part of uninterruptible power supplies for critical industries where a continuous supply of power is vital. These systems are often used for load leveling and power quality control in place of batteries because of their smaller size and their less demanding maintenance requirements.

Direct grid storage applications for flywheels include a 20 MW, 150 MWh flywheel storage plant that was installed at Stephentown, New York where it provides frequency regulation services for the New York grid operator.[3] The storage facility is made up of 200 flywheels and completes between 3000 and 5000 full-depth discharges each year. The unit can supply 20 MW for 15 min. Similarly sized units have been built to provide power over a much longer period, measured in hours rather than minutes. A schematic of a storage facility with multiple flywheels.

Flywheel storage has also been combined with wind energy generation. This combination allows the wind plant to store energy when it is not needed by the grid and then supply it as demand rises. It is one of a number of ways that different energy storage technologies are being combined with renewable energy to provide a more reliable and therefore more easily dispatchable and more valuable electricity source from intermittent energy sources.

[3]http://beaconpower.com/stephentown-new-york/.

CHAPTER 7

Super-Capacitors

Capacitors are electrical or electronic devices that store electrical energy directly in the form of electrostatic charge. The simplest capacitor comprises two metal plates separated by a small air gap. When a voltage is applied across the device, the plates become statically charged. If the voltage is removed, the static charge will remain until a short circuit is applied between the plates. The amount of charge on each plate builds up until there is sufficient to create an electric field that exactly balances that created by the voltage.

Since the capacitor is essentially a break in the continuity of the circuit, the device will not allow any direct current (DC) to pass. However it will transmit an alternating current (AC). The ease with which AC current can pass through it depends on the frequency; the higher the frequency, the larger the amplitude of the signal that passes through the capacitor. Small capacitors are one of the three primary components used in electrical and electronic circuits. Together with resistors and inductors they can provide circuity for a range of different functions that are required in most electronic devices. (Both inductors and capacitors can be thought of frequency variable resistors; the resistance of a capacitor at DC is infinite but it falls as the AC frequency rises. An inductor has low resistance to DC but its resistance rises with the AC frequency.)

The unit of capacitance is the Farad. A capacitor of 1 F will have a voltage of 1 V across its plates when it has stored 1 C of electric charge. The Farad is an unreasonably large capacitance for most practical applications and capacitors used in electrical and electronic circuitry are normally measured in micro-Farad or smaller. Modern energy storage capacitors, in contrast, can be measured in the 100–10,000 F range of storage capacities.

The important characteristic of the capacitor for energy storage is its ability to store electrostatic charge directly. For this to be useful for grid supply applications, the storage capacity must be very large.

Power System Energy Storage Technologies. DOI: https://doi.org/10.1016/B978-0-12-812902-9.00007-9

The largest conventional capacitors are called electrolytic capacitors because they have an electrolyte in liquid, gel, or solid form acting as one of their electrodes. These provide large capacitances for electrical and electronic circuitry but are still too small for most energy storage applications. Since the 1970s a new type of capacitor called an electro-chemical capacitor has been developed. These also involve electrolytes. However electrochemical capacitors utilize structures and concepts that are similar to some of those involved in batteries. Several different types of electrochemical capacitor have been discovered. Today the devices are sold commercially under names such as super-capacitors, ultra-capacitors, and electric double-layer capacitors.

As already noted, a key advantage of these devices from an energy storage perspective is that they store electricity in the form of electrical charge directly, rather than converting it into another form of energy. This means that the electrical energy, the electrostatic charge contained in the super-capacitor, is available immediately and this makes them extremely fast acting. As storage devices, super-capacitors can absorb and deliver power more quickly than a battery and they can be charged and discharged more times too. However they are limited to relatively small absolute energy storage capacity; the largest can store up to 8 kWh. Current super-capacitors have a storage energy density of roughly one-tenth that of a lithium ion battery. The voltage from a super-capacitor falls as its charge falls, too, whereas that of a battery will remain broadly the same through most of its discharge cycle. This affects the way in which each can be used.

ENERGY STORAGE CAPACITOR PRINCIPLES

As outlined earlier, a simple electrostatic capacitor comprises two plates with an air gap between them (see Fig. 7.1). When a voltage is applied to the plates, charge builds upon them in order to neutralize the voltage by creating an equal and opposite static-charge voltage across the plates. The charge will continue to build as the voltage is increased until it is high enough to cause air to breakdown and start conducting electricity between the plates. At this point the capacitor fails.

The amount of charge that the capacitor will hold can be increased by placing a dielectric material between the plates. This

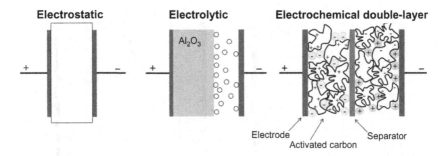

Figure 7.1 Schematic showing different types of capacitor. Source: Wikimedia.

reduces the strength of the field between the plates, allowing more charge to buildup before breakdown occurs. In essence the dielectric itself becomes polarized when an electric field is applied to the capacitor plates and this allows the device to store more charge than the plates alone would be able to store because some is stored within the dielectric. For conventional small capacitors the dielectric material is a solid.

An electrolytic capacitor, the largest of the conventional capacitors, is similar to this in that it has a dielectric material between the capacitor plates. However in this case the dielectric material is a rough oxide layer formed by "anodizing," the surface of a material such as aluminum in an electrolyte. This rough layer, which forms one plate of the capacitor, has a very large surface area. In early electrolytic capacitors the second electrode was created by immersing the plate in a liquid electrolyte into which an electrical contact was applied. The liquid was later replaced by a dry gel to create a "dry" electrolytic capacitor. The liquid or gel is necessary because a solid electrode would not be able to come into contact with the all the irregular surface of the oxide layer. This type of capacitor functions in a similar way to a double plate capacitor with a dielectric between the plates but the large surface area of the oxide layer allows it to store a large capacity of charge. The principle behind the electrolytic capacitor was first explored in the 1920s. A simple electrolytic capacitor is shown schematically in Fig. 7.1.

An electrochemical capacitor extends this concept further still. It has electrodes that act as the plates of the device and these are immersed in an electrolyte that contains ions that can move between the plates. There is also a semipermeable membrane within the

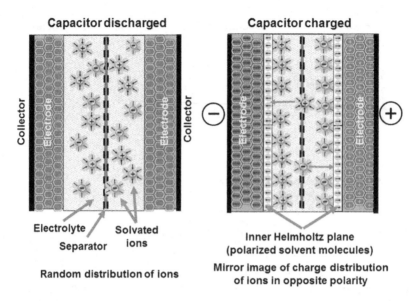

Capacitor discharged

Capacitor charged

Electrolyte / Solvated ions

Separator

Random distribution of ions

Inner Helmholtz plane
(polarized solvent molecules)

Mirror image of charge distribution
of ions in opposite polarity

Figure 7.2 Diagram of a symmetrical electrochemical capacitor in discharged and charged states.
Source: Wikimedia.

electrolyte that separates the plates. The job of this membrane is to ensure that the plates cannot come into contact and create a short circuit. The first devices of this type were invented in the United States in the 1950s but it was not until the 1970s and 1980s that commercial version began to appear. A simplified version of an electrochemical capacitor is also shown schematically in Fig. 7.1.

When a charge is applied to the capacitor, it causes the usual charge buildup. However in this case the charge on each plate is neutralized by an opposing layer of charged ions, called a Helmholtz layer, from the electrolyte. This creates a double layer of charge at each plate, effectively leading to two capacitive charged layers, which massively increases the amount of charge that the unit can hold. A more detailed view of an electrochemical capacitor is shown in Fig. 7.2.

The simplest electrochemical capacitors use a water-based electrolyte and this restricts the voltage that each can support to around 1.23 V. If the voltage strays above this, electrolysis of the water begins to take place making the structure unstable.[1] The voltage restriction limits the

[1]The minimum voltage for the electrolysis of water is 1.23 V. However a higher voltage often has to be applied. Some aqueous electrolyte super-capacitors are rated at 2.1–2.3 V.

amount of charge that the capacitor can hold because the quantity of energy stored is proportional to the square of the voltage, but otherwise these capacitors have excellent characteristics. Basic capacitors of this type (known as symmetrical capacitors) have two identical electrodes usually made from some form of amorphous carbon. By varying the construction of one electrode, it is possible to increase the amount of energy the device can hold. Devices of this type are called asymmetrical capacitors because the two plates hold different amounts of charge.

A second type of electrochemical capacitor uses an organic electrolyte and this allows it to support a voltage of up to 2.7 V and so contain a higher charge. Both symmetrical and asymmetrical versions of these capacitors are available too. Since single capacitors of either type can only support a relatively low voltage (in grid terms), electrochemical capacitors are usually stacked in series in order to allow higher voltages to be exploited.

The amount of charge that an electrochemical capacitor can store can be increased if there are molecules in the electrolyte that can become charged and absorbed onto the surface of the electrode when a voltage is applied across the capacitor. Some ionic species might also be absorbed into the structure of the electrode when a voltage is applied. The additional capacitance that this can add to the device is called pseudocapacitance and it acts alongside the normal electrode double-layer capacitance. Super-capacitors with this ability can provide even higher storage capacities than the more conventional electro-chemical capacitors.

Key to the functioning of a super-capacitor is the structure of each electrode. For a conventional super-capacitor these electrodes are normally made by applying a porous coating, usually of some form or carbon, to a metal electrode plate. The amount of charge that the electrode can hold depends on the surface area of the electrode since this is where the charge accumulates, so this is made a large as possible by using some form of carbon that has been "activated" to increase its surface area. Activated carbon has been widely used, as well as more modern forms including carbon aerogels, graphene, and carbon nano-tubes. Pseudocapacitors have electrodes made of metal oxides, typically manganese oxide or ruthenium oxide because these can support the sort of reversible reactions that lead to a large charge storage capacity. The oxide layer is porous to provide high surface area.

The electrolyte in a super-capacitor is a solvent with a material dissolved in it which dissociates into positively charged ions (cations) and negatively charged ions (anions). It is these ions that migrate to the electrodes when a voltage is applied to the device and form the charged double layers where the capacitor stores energy. Many super-capacitors use water as the solvent with acids, alkalis, or salts added to create the necessary cations and anions. This limits the maximum voltage that the device can support. If this is exceeded, then internal reactions start to take place. More recently organic solvents have also been used in super-capacitors. These can withstand higher voltages before breaking down but have lower conductivity.

PERFORMANCE CHARACTERISTICS

Super-capacitors are unconventional devices and they have some odd properties. For example, the capacitance of a super-capacitor depends upon how it is measured. For energy storage, the key storage capacity is the DC storage capacity and it is this capacitance that will be the most significant. However capacitance is normally measured using an AC signal. This leads to a much lower capacitance value that found under DC conditions. Even at a frequency as low as 10 Hz, the capacitance will have dropped to 20% of less of its nominal DC value. The strong frequency dependence is a result of the speed at which the ions that create the stored charge can move. The fact that this frequency dependence exists can also affect the operational energy storage capacity of a super-capacitor in situations where it is charged and discharged extremely rapidly. For longer term storage applications, the full DC storage capacity will be available.

On the other hand, electrochemical capacitors can be cycled for tens of thousands of times without degradation provided the voltage across them is kept below the maximum so that no internal reaction takes place. This compares favorably with batteries that have a much more limited cycle life.

Once a super-capacitor is charged, it will lose charge slowly through leakage of charge across the electrolyte barrier from one electrode to the other. (Similar processes limit the long-term storage capability of most batteries.) The rate of charge leakage in water-based electrochemical cells is similar to that of a lead-acid battery. Leakage levels are lower with organic-based electrolytes.

Leakage will reduce long-term storage life but with fast cycling, round trip efficiency can be 95% or higher. In addition, capacitors can normally discharge their energy very rapidly without damage so long as excessive internal heating is avoided. This heating is caused by the internal resistance of the super-capacitor. Heat generated within the device must be dissipated; overheating will limit the life of the device. The specific energy density of a storage device measures the amount of energy it stores for each unit of its mass. The specific energy density of a super-capacitor is typically between 1 and 5 Wh/kg for symmetrical capacitors and up to 20 Wh/kg for asymmetrical capacitors. In comparison a lead-acid battery has an energy density of up to 42 Wh/kg while a lithium ion battery as a specific energy density of up to 250 Wh/kg.

Another parameter, specific power, provides a measure of the speed at which power can be delivered or absorbed. The specific power of a super-capacitor is up to 15 kW/kg. This is 10−100 times greater than the specific power of most battery systems. Response time is fast at 5 ms.

Units can be designed with the capability to deliver power levels of between 1 kW and 5 MW but actual energy storage capacity is generally relatively low at between 1 and 10 kWh. When power is withdrawn from a capacitor, the voltage falls so sophisticated DC−AC conversion systems are needed to maintain a constant voltage output as the super-capacitor voltage falls.

APPLICATIONS

Electrochemical capacitors have been used both for energy storage and for braking energy recovery systems in automotive applications. For grid use they are best suited to backup or fast reaction grid support, offering a similar performance to flywheels. Although capacitors are not yet widely deployed for grid support, they have been tested in a number of configurations. These include adding rapid response storage to small distributed generation grids or microgrids where they can provide fast reacting grid support when the output from intermittent renewable resources suddenly falls and before a backup engine-based system can take over. Capacitors are also being tested for high voltage grid support services. The fast response time of the super-capacitor means that it can absorb spikes of transient power extremely effectively.

There may also be a role for super-capacitors as short-term storage devices coupled with renewable energy sources. However their small absolute storage capacities limit their use in many grid storage applications.

CHAPTER 8

Hydrogen Energy Storage

Most of the energy storage technologies discussed in this volume are designed specifically for electrical energy storage or are adapted for this particular application. Hydrogen energy storage is slightly different because it offers a general purpose means of storing energy and that energy can then be used in a range of different ways. The gas is attractive because it is a low-carbon energy source since its use does not lead to carbon dioxide emissions during use (although its production might release carbon dioxide). This makes hydrogen energy storage potentially one of the most important means of storing energy available.

To make it feasible, however, requires the development of what has become known as the hydrogen economy. This is a fuel economy that is built around hydrogen instead of fossil fuels. Hydrogen does not occur naturally in large amounts on the earth but it can be produced from a number of sources and is widely used in the chemical industry. Most hydrogen today is made from natural gas but some is also produced from water and it can be made from chemicals such as alcohols too. The latter can be produced naturally. It costs energy to make hydrogen from any of these sources: in essence this energy is stored in the hydrogen and can then be released later. However whereas the original source of the energy may have had limited applications, hydrogen itself is extremely versatile.

The hydrogen economy does not exist today and building it will be a massive infrastructure project. There will have to be large-scale means of producing hydrogen and then methods of distributing it in different ways to suit different users. Some might be shipped in pipelines while elsewhere it will be delivered by tanker to local distribution centers. Storage, itself, is a problem because hydrogen is a gas down to very low temperatures so finding a way of storing it in a concentrated

Power System Energy Storage Technologies. DOI: https://doi.org/10.1016/B978-0-12-812902-9.00008-0

form may require complex technological solutions. Storage in steel cylinders under high pressure will be the default until other, better methods are found. Fortunately hydrogen has a high energy density and this is an advantage.

Most hydrogen today is produced chemically from fossil fuel but in order to create a carbon free economy, as well as provide the massive volumes required, the gas would probably have to be produced primarily by the electrolysis of water. Water is one of the most abundant compounds on earth but the production of hydrogen using electricity is relatively inefficient today. If hydrogen storage is to have a future, then low-cost sources of renewable electricity will be required, surplus power that can be used to produce the hydrogen needed for all the other applications. Whether it is more cost effective to produce hydrogen in this way rather than to use the electrical power directly will depend upon many factors and the answer today is not clear.

Meanwhile, hydrogen is attractive as an energy storage medium because of its versatility. Its production from water is simple, if relatively costly today. Once made, the gas can be used directly in a variety of combustion plants, including gas turbines and gas-fired steam turbine plants, reciprocating engines and, most importantly, the fuel cell. When it burns in air, the only combustion product is water. In addition hydrogen can be used in fuel cell−powered vehicles, although these remain costly compared to other emission free technologies. Moreover it can easily be used for domestic and commercial heating and for domestic cooking, a direct replacement for natural gas.

While most visions of a hydrogen economy envisage large-scale production plants, it might be more effective to have many more small, distributed production facilities. One of the options that are being explored is to combine an intermittent renewable source of power such as wind of solar power with a hydrogen production plant. Surplus power when the wind blows or the sun shines but demand is low can be converted into hydrogen which can then—in theory anyway—be delivered into the local hydrogen infrastructure.

Many questions about the hydrogen economy and about hydrogen energy storage remain to be answered. In the meantime it is being promoted as a complement to other forms of energy storage discussed here.

THE PRINCIPLES OF HYDROGEN ENERGY STORAGE

In essence the hydrogen energy concept, as applied to electrical energy storage, is extremely simple. Surplus or off-peak electrical power is used to make hydrogen and this hydrogen in stored. Then, when electric power is required, the stored hydrogen is used as fuel in a power plant. While the concept is simple and the technology to implement it is available today, there are problems in each of the three stages on the process, production, storage, and power generation, that will need to be solved in order to make this a large-scale solution to energy storage and a low-carbon economy. A schematic of a simple hydrogen energy storage system based on renewable energy sources is shown in Fig. 8.1.

Hydrogen production is carried out on an industrial scale today, normally by a process called reforming of hydrocarbons such as natural gas that is produced by the oil and gas industry. This is a carbon intensive production method with carbon dioxide as one of the main by-products. During the process, the greenhouse gas is usually released into the atmosphere, adding to the global atmospheric load. The annual production of hydrogen is between 200 and 300 bn m^3. Virtually all of this is produced by reforming. However in a hydrogen economy the quantity of hydrogen required would be many times

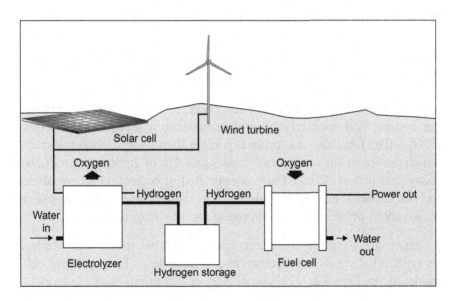

Figure 8.1 Simple hydrogen energy storage system.

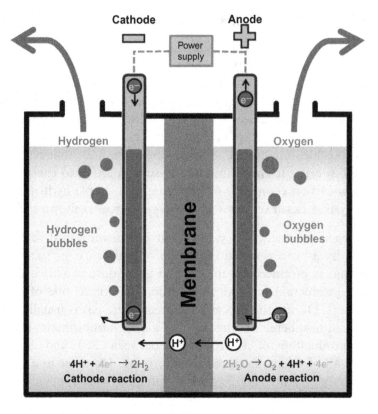

Figure 8.2 Schematic of an electrolysis cell. Source: US Office of Energy Efficiency and Renewable Energy.[1]

this volume. As a guide to the amount that might be needed, global natural gas consumption in 2016 was over 3500 bn m^3.[2]

Reforming is economical but has no place in an electric energy storage system. This must rely on another production method, the electrolysis of water. Electrolysis is more expensive than hydrocarbon reforming, which is why it only accounts for around 4% of hydrogen production today, but it is available. There is estimated to be around 8 GW of electrolysis capacity around the globe. This would have to be ramped up massively to provide enough hydrogen for a hydrogen economy.

There are a variety of electrolyzers available today but whatever its type, all electrolyzers operate in a similar way. This is shown schematically in Fig. 8.2. The electrolysis cell contains two electrodes, a

[1]https://energy.gov/eere/fuelcells/hydrogen-production-electrolysis.
[2]BP Statistical Review of World Energy, 2017.

negatively charged electrode called the cathode and a positively charged electrode called the anode. These are immersed into an electrolyte which in this case is acidic, and so contains a high concentration of hydrogen ions (H^+). When a sufficiently high voltage is applied across the electrodes, and the minimum under ideal conditions is 1.23 V (in practice a higher voltage is usually required), then the positively charged hydrogen ions in the electrolyte migrate to the cathode where they combine, with the addition of electrons from the electrode, to produce gaseous hydrogen. At the same time water molecules at the positive electrode react to form oxygen while releasing more hydrogen ions into the electrolyte. Another variety of electrolyzer, called an alkaline electrolyzer, utilizes an alkaline electrolyte that contains a high concentration of hydroxide ions (OH^-). In this case, these migrate through the electrolyte to the cathode of the cell where they react to produce oxygen while more hydroxide ions are produced from water at the cathode, with the release of hydrogen gas.

The electrolysis of water using electrical power has been carried out industrially for many years. The main system in use in the 21st century is the alkaline electrolyzer, exploited most successfully by the Norwegian utility Norsk Hydro. Large-scale electrolyzers have been built that are capable of handling inputs of 100 MW for hydrogen generation and the product is around 99.8% pure. Conversion efficiency is 90%. Alternatives to the alkaline electrolyzer are the proton exchange membrane (PEM) electrolyzer that is currently being developed and could potentially achieve 94% efficiency, but with the need for a platinum catalyst. The PEM electrolyzer is similar to a PEM fuel cell, with the membrane forming the electrolyte. High-temperature ceramic electrolyzers are also under development. These use solid oxides as the electrolyte, similar to those used in solid oxide fuel cells. While large-scale electrolyzers may be needed in the future, the immediate application is likely to be for small-scale electrolyzers that might be coupled with renewable energy systems to help manage the energy production from these generators.

Once produced, hydrogen must be stored. While the gas has a high mass energy density, it is very light and has a low-volume energy density so it must be compressed or stored in a concentrated state to achieve a high-volume energy density similar to other common fuels.

For power generation applications, storage under pressure in steel or composite tanks is probably the favored method. Pressures of 350−700 bar are common.

Simple high-temperature compression of hydrogen may be suitable for many small-scale hydrogen storage systems based on renewable energy. However this is relatively simple technology and more advanced storage methods are expected to be able to achieve much greater energy density. The best alternative physical methods of storage involve cooling the gas using a cryogenic cooling system. Cryogenic cooling and compression can achieve much higher energy density, depending upon the temperature to which the gas is cooled, the lower the temperature the better. With this type of system the hydrogen is still in gaseous form. However if it is cooled to a low enough temperature, below $10°K$ ($-263°C$), the gas will liquefy and it can then be stored in liquid form. Both these low-temperature approaches require the continuous cooling to maintain the low temperature and this will add to the operational costs.

There are a number of alternative methods of storing hydrogen that are being evaluated. Some of these involve the adsorption of hydrogen onto the surface of a high-surface-area material while others involve its intercalation into the structure of a material. The latter usually involves the formation of a metal hydride where the hydrogen is weakly bonded within the structure of the absorbent and can be released easy, for example, by heating. Other methods include the formation of chemical compounds in which the hydrogen is, again, weakly bonded and can be released with a low energy cost.

An alternative that can provide very high-volume storage capability is underground storage. Underground storage caverns created from salt domes have been used to store natural gas for many years and some chemical companies have used similar structures for hydrogen storage. However this depends on suitable underground geological features being available. Storage facilities of this size and type are likely to be essential to support a hydrogen economy.

One manufactured and stored, hydrogen can be converted back into electricity in a number of ways. It can be burned like natural gas, although the combustion temperature is generally higher than for the former. However most gas turbines, piston engines, and gas-fired boilers can easily be adapted for its use. A common method of reducing the combustion temperature is called exhaust gas recycling. This involves some of the gas from the exhaust of the plant being mixed with air that is fed into the combustion engine. Since the exhaust gas has already taken part in the combustion process, the concentration of oxygen that it contains is reduced. This affects the reaction dynamics and helps reduce the combustion temperature.

Adapting combustion engines for power generation from hydrogen can be achieved easily today and offers the simplest route for small-scale adoption. For the future, however, the most important method of generating power from hydrogen is likely to be with the hydrogen fuel cell. A number of these have already been developed and some are available commercially. These include phosphoric acid fuel cells, PEM fuel cells, molten carbonate fuel cells, and solid oxide fuel cells. What these have in common is that they use an electrochemical cell to produce electricity from hydrogen and oxygen (from air). Efficiencies vary but the best of them can achieve up to 60% efficiency in a simple fuel cell. With more complex fuel cell systems it may be possible to achieve higher efficiency still, perhaps as high as 70% or 75%.

PERFORMANCE CHARACTERISTICS

Energy storage systems are usually characterized by properties such as their storage capacities, their response times, and the rate at which they can deliver power to the grid. Hydrogen cannot be characterized simply in this way because the actual characteristics depend on the method used to convert the gas back into electrical power.

For example, if the gas is used as fuel in a conventional combustion power plant then its performance characteristics will be defined by the type of combustion engine used in the plant. A reciprocating gas engine burning hydrogen can respond relatively quickly to a demand for power but start-up time may be seconds if it is not already turning over. It will respond to changes in demand easily but the maximum power for this sort of device is limited to around 10 MW and most are much smaller. Gas turbine systems can provide higher power but are generally slightly more complex. Their response times are likely to be slower than those of a reciprocating engine.

Fuel cells probably offer the best performance in terms of response. They can be started relatively quickly and can respond well to changes in demand too. Efficiency is relatively flat from 0% to 100% output whereas that of a combustion engine will fall off as the output falls. Fuel cells are still expensive and single fuel cells are normally less than 1 MW in capacity. For large capacity plants, multiple fuel cells must be coupled together. However none of the methods of converting hydrogen back into electrical power can be considered very fast acting, when compared to systems like batteries, flywheels, or capacitors. This means that hydrogen energy storage is more suited to energy arbitrage and long-term storage that for grid support.

On the other hand the energy storage capacity of a hydrogen energy storage system can be extremely large and is only limited by the capacity of the vessel used to store the gas. If high energy density storage is not at issue, hydrogen can be stored in underground caverns such as those created from salt domes. These have been used for decades to store natural gas and some industrial companies have also stored hydrogen in similar facilities without problem. The gas can be stored under pressure to increase the capacity and this is likely to form a part of any future hydrogen economy infrastructure. With such facilities, hydrogen storage can be considered a large-scale energy storage system, similar to pumped storage hydropower.

Hydrogen storage is feasible today but its main drawback is the round cycle efficiency. With an electrolyzer operating at 90% efficiency and a power plant converting it back into electricity with perhaps 60% efficiency, the best round trip efficiency that can be expected is 54%, much lower than the other storage systems that have already been considered in previous chapters. Such low efficiency may be tolerable in a renewable energy storage system such as a wind-hydrogen storage unit where the wind energy must otherwise be shed. It is unlikely to be considered sufficiently efficient for generation from off-peak grid power in most other circumstances if there is an alternative available.

APPLICATIONS OF HYDROGEN ENERGY STORAGE

Once hydrogen has been produced from an energy system, it can be reused in a large number of ways. One is to use it in industrial processes that currently take hydrogen from the reforming of hydrocarbons. In a low-carbon world, industrial hydrogen will of necessity come from the electrolysis of water using cheap electrical power.

An important focus of research today is into the use of hydrogen as a fuel for vehicles. Fuel cell cars are already available commercially but they are more expensive than other types of environmentally benign vehicle such as battery powered vehicles. One of the key challenges here is to find a cheap and effective means of providing a vehicle with a large charge of hydrogen so that it has a useful range. Some of the chemical methods of storing hydrogen discussed earlier may provide this in the future.

For the electricity utility industry, the future of hydrogen is less clear. There are schemes that are testing or have tested the use of hydrogen storage to capture energy from renewable plants, mainly wind farms, when the power would otherwise be shed. However this then requires a means to convert the hydrogen back into electricity at the wind farm. Whether this is cost effective compared to, say, using a large-scale battery system has yet to be established.

If the hydrogen economy should come into existence, that could change the outlook dramatically. Hydrogen would then be available widely and there would be storage facilities everywhere, too. This would make hydrogen production and storage from power plants much cheaper and easier to integrate into the wider operation of the grid. A generator might, for example, make hydrogen in one region and then sell it in another in the same way as natural gas and power can be bought and sold without the need to physically exchange the same bottle of the gas. That scenario is, however, still a long way off and may never become a reality.

CHAPTER 9

The Environmental Impact of Energy Storage Technologies

The different types of storage plant discussed in this volume offer a diverse range of systems for storing energy. As technologies they each rely on a different principle. Some of these are technically complex and involve the use of advanced materials. Superconducting magnetic energy storage (SMES) would be a typical example. Others, such as pumped storage hydropower, depend on well-established and well-understood technologies that have been perfected over more than a century.

From one perspective, the environmental impact of each of these storage systems will depend on the precise nature of the technology it exploits. For example, construction of a pumped storage hydropower plant requires massive civil works in order to provide the storage reservoirs and powerhouse for the plant. This may cause significant local disruption and environmental upheaval. However once the plant has been built, it should cause little further disruption if it acts only as an energy storage plant. A compressed air energy storage plant in which air is stored in an underground cavern will have relatively little local impact during construction. On the other hand, depending on the source of stored energy and the way in which it is operated, it may release carbon dioxide into the atmosphere. Battery systems are self-contained and factory constructed but they often contain toxic materials. Small systems such as flywheels or super-capacitors will have little local impact, unless there is an accident.

From another perspective, however, all energy storage technologies can be considered as a means of making the operation of an electricity supply system more economical and cleaner. The load shifting that they provide can reduce carbon emissions by allowing the most efficient power plants, in terms of emissions, to provide the maximum amount of energy. In the longer term, these technologies will also be a vital cog in a power

Power System Energy Storage Technologies. DOI: https://doi.org/10.1016/B978-0-12-812902-9.00009-2

supply system that is essentially carbon free because they can be used to convert intermittent renewable sources of power into reliable and dispatchable power plants. From this point of view, therefore, they are all environmentally advantageous. Balancing the environmental pros and cons of energy storage will usually end up in their favor.

TECHNOLOGY-SPECIFIC ENVIRONMENTAL CONSIDERATIONS

The technologies that underpin each of the energy storage systems discussed in this volume differ and each has its own environmental implications. Some of these have been discussed in detail in other volumes of the series but the main features and considerations entailed by each will be outlined here. One key consideration that applies to all of the technologies is the source of the energy that is being stored. If the power comes from renewable sources such as wind, solar, or hydropower, then it can be considered sustainable and essentially emission free. However if it comes from plants burning fossil fuel, or from nuclear power plants, then the environmental implications of the operation of these power plants must be taken into account. Stored energy from a coal-fired power plant still has a burden of environmental emissions attached to its production and use, even if the latter has been delayed by storing the energy for a period in an energy storage facility. On the other hand, the energy storage plant itself is usually neutral and as a result of its own operation will not—in most cases—produce any additional emissions.

Pumped storage hydropower: A pumped storage hydropower plant has two reservoirs, often with a dam for each, and involves large-scale civil engineering work to construct. Many of these plants are very large, often more than 1000 MW in generating capacity and the dams are massive physical structures. These will inevitably have an impact on the local environment, both during construction and later in operation. The construction of a dam and the impounding of a reservoir will create an enormous additional physical load on the local geological substructures and this can lead to small seismic events. Meanwhile the creation of a reservoir will inundate a large area of land, destroying all the wildlife and plants; any buildings or other physical structures in the area covered by the reservoir will be underwater for at least part of the time.[1] If the plant is on a river and can act as a conventional

[1]Some reservoirs are built in mine workings or other man-made structures. In this case the disruption in converting it into a pumped storage facility will be lessened.

hydropower plant too, then water flows will be affected downstream. On the other hand a plant designed solely for energy storage will have limited impact once it has been built and its reservoirs charged. These plants are among the largest energy storage plants available and they are capable of providing massive levels of grid support and energy arbitrage services.

Compressed air energy storage: A compressed air energy storage facility is much like a gas turbine power plant, but with a store for compressed air attached. The environmental effect of this will depend on how the plant is operated. If the plant takes power from the grid to drive its compressor and store compressed air, then uses the compressed air to drive an air turbine directly, the overall impact of the plant will be limited. However there is an option to operate these plants in a different way, burning fuel such as natural gas to increase the energy output from the plant when it is supplying power using the compressed air from its energy store. In this case there will be carbon dioxide emissions that must be taken into account, as well as other emissions associated with a fossil fuel combustion plant. The other main consideration is the compressed air store. For larger plants this is likely to be some form of underground cavern. There may be local security issues associated with the operation of this but except in exceptional cases these should be limited.

Battery energy storage plants: There are a large number of different battery energy storage technologies and they differ in their environmental implications. The main environmental issue with battery systems relates to the materials from which they are made. These are often exotic and some are hazardous. For example, several battery systems use cadmium that can affect both animals and humans if ingested. Furthermore, if it enters the body, it will remain there for a long time. Lead is another material, common in lead-acid batteries, that can be hazardous if not managed carefully. Lithium, which is used in lithium ion and lithium hydride batteries, is extremely reactive when exposed to air and battery systems containing it must be designed with this in mind. Lithium batteries are also capable of catastrophic failure and safety circuits are required to ensure that their operating conditions are carefully controlled. Sodium sulfur batteries are also potentially hazardous due to the presence of liquid sodium. A small number of explosions have been reported. The other issue with battery systems is their fate when they are withdrawn from service. Lead- and cadmium-containing batteries will normally be recycled but there is little recycling of lithium batteries carried out today, although these battery systems are growing strongly in popularity.

Superconducting magnetic energy storage: A SMES storage device is a high technology storage system. It relies on a coil made from metal alloys or special ceramics that must be cooled to extremely low temperatures in order to function. The low-temperature refrigeration system can be hazardous if not managed carefully. The systems create a very powerful magnetic field that must be carefully shielded. The magnetic field also puts elements of the structure under great stress so components must be engineered to withstand these forces. However there are no major environmental issues associated with this technology that would not be associated with electrical, magnetic, and low-temperature systems elsewhere.

Flywheels: Flywheel energy systems are relatively small, mechanical-based energy storage devices. They contain motors and generators as well as the flywheel itself. All these elements will generate some noise but this can be limited with good sound insulation. The main hazard associated with flywheels is that of catastrophic failure of the flywheel itself due to the massive centrifugal forces it experiences when operating. To counter this flywheel systems must be housed in massive enclosures that can contain the parts of the device, which will be launched at massive speed, should it come apart. Other than that there are no obvious environmental hazards excepting than those associated with all high-speed mechanical machines, for which care should be taken when in operation.

Super-capacitors: A super-capacitor is a relatively compact, small-scale energy storage system based on advanced electrochemical technologies. The device can store a large amount of electrical charge in a relatively small space, and if this is discharged quickly it has the potential to cause significant damage, both externally and internally. The materials from which super-capacitors are manufactured are relatively commonplace in their composition, if not in their construction. In most cases they are not hazardous of themselves although because the devices are still under development, some future versions may contain more exotic and hazardous elements. There are arguments that super-capacitors are preferable to the more common lithium ion batteries because of environmental issues with the latter, suggesting that these devices are considered environmentally benign from most perspectives.

Hydrogen energy storage: The use of hydrogen as an energy storage medium involves three stages, hydrogen production, hydrogen storage, and hydrogen use. Production using surplus electrical power is a simple and straightforward process with few hazards except those associated with the product gases themselves, hydrogen and the by-product, oxygen. However the challenges associated with storage are greater. The issues here are similar to those associated with the handling and storage of natural gas, with the exception that accidental release of hydrogen into the atmosphere, provided it dissipated quickly, has no significant environmental impact. There is no danger to humans or animals from contact with the gas. However it is explosive so care has to be taken that it does not build up in enclosed spaces. When hydrogen is to be used to generate electric power, this will be by reaction with oxygen, either in a combustion plant or in a fuel cell. In both cases the oxygen will normally come from air and the reaction product will be water, usually in gaseous form. This is considered environmentally benign.

THE ENVIRONMENTAL IMPORTANCE OF ENERGY STORAGE

While there are a range of environmental issues associated with the energy storage technologies, noted earlier, the main environmental impact of storage is advantageous from an environmental point of view. There are two primary applications for energy storage systems, to support the grid by providing grid stability services and to provide energy arbitrage or long-term storage. Both make grids more stable, cheaper, and easier to operate, and over the longer term they will allow cleaner technologies to provide most if not all of the world's electrical energy. As such they are one of the key technologies needed to help combat global warming.

Grid stability can be enhanced at all levels with energy storage devices but the small, fast-acting devices are perhaps the most important here as they can provide improved stability at a local level. For example, a SMES system can be used to support and stabilize a long spur distribution line, while a flywheel system can provide power conditioning services to a sensitive computer installation. These fast-acting systems can intervene within the space of a grid cycle to correct instability such as might be caused by a lightning strike or the sudden failure of a grid energy source. The type of service that these devices can offer is not available from the passive devices that have commonly been used to maintain grid stability in the past. A stable grid operates more efficiently and therefore in a more environmentally friendly way.

The more far-reaching and long-term advantage of energy storage, however, is to provide forms of load leveling or energy arbitrage. This can be both short-term and long-term, small-scale and large-scale, and there are storage technologies that can provide any combination of these requirements.

The most significant application of this type is in the support of intermittent renewable energy sources such as a wind and solar power. Wind power is the most unpredictable renewable source and this makes wind energy difficult for grid controllers to dispatch. However if wind power can be combined with significant energy storage capacity, then wind energy that is available when the wind blows can be used, or stored, as necessary and then when the wind stops blowing, stored energy can be drawn on until it starts blowing again. Solar power is more predictable but it is not available for 24 h each day. Storage can make it so. Tidal power is also intermittent and can be treated in the same way. Of course much depends on the available storage capacity compared to demand but in principle this approach can be used to enable most if not all of grid power to be supplied by such intermittent sources.

In spite of these advantages, there is still relatively little energy storage capacity on most of the world's grids. The main argument against greater capacity has been the cost of the technologies. It is expensive to build energy storage capacity and grid tariffs do not always reward storage in a way that encourages its construction. However, attitudes are changing slowly and it is likely that more storage will be built in the future.

CHAPTER 10

The Cost and Economics of Energy Storage

Evaluating the cost of energy storage is difficult because the value of the energy a storage plant can hold depends critically on the prevailing conditions under which it operates. A storage plant operating in load shifting mode will buy cheap, off-peak power and then sell it back to the grid at times of high or peak demand. The value of the energy it holds is then found in the difference between the cost of buying the power and the rewards gained from selling it. Other storage plants may provide grid stability or power conditioning services. The value in this case is determined by the losses that might be incurred if the grid became unstable or power quality dropped.

The broader value of energy storage plants as a means of enabling cleaner sources of electricity to supply large quantities of power is even more difficult to establish. These plants may store power over both the short term and the long term. Estimating the value of this depends on putting a value on the damage to the global environment caused by the alternative less environmentally friendly power plants that can be replaced if energy storage is used. Making such estimates is feasible but often contentious. And while it may be difficult to place a value on energy storage, the cost of building energy storage plants is generally high. This makes decisions to build them difficult to make. It is important to put this high cost into context if the case for energy storage is to be made compelling.

There has, recently, been a systematic attempt to cost energy storage using an economic analytical approach called the levelized cost model. This has been used for many decades to establish the cost of electricity from different types of power plant. Now, a similar approach has been applied to storage plants.

Power System Energy Storage Technologies. DOI: https://doi.org/10.1016/B978-0-12-812902-9.00010-9

LEVELIZED COST MODEL

The levelized cost approach is most easily understood when considering a power-generating plant. The cost of electricity from a power plant of any type depends on a range of factors. First there is the cost of building the power station and buying all the components needed for its construction. In addition, many power projects today are financed using loans so there will also be a cost associated with paying back the loan, with interest. Then there is the cost of operating and maintaining the plant over its lifetime. Finally the overall cost equation should include the cost of decommissioning the power station once it is removed from service.

It would be possible to add up all these cost elements to provide a total cost of building and running the power station over its lifetime, including the cost of decommissioning, and then dividing this total by the total number of units of electricity that the power station actually produced over its lifetime. The result would be the real lifetime cost of electricity from the plant. Unfortunately such as calculation could only be completed once the power station was no longer in service. From a practical point of view, this would not be of much use. The point in time at which the cost of electricity calculation of this type is most needed is before the power station is built. This is when a decision is made to build a particular type of power plant, based normally on the technology that will offer the least cost electricity over its lifetime.

In order to get around this problem, economists have devised a model that provides an estimate of the lifetime cost of electricity before the station is built. Of course, since the plant does not yet exist, the model requires that a large number of assumptions be made. In order to make this model as useful as possible, all future costs are also converted to the equivalent cost today by using a parameter known as the discount rate. The discount rate is almost the same as the interest rate and relates to the way in which the value of one unit of currency falls (most usually, but it could rise) in the future. This allows, for example, the cost of replacement of a plant component 20 years into the future to be converted into an equivalent cost today. The discount rate can also be applied the cost of electricity from the power plant in 20 years time.

The economic model is called the levelized cost of electricity (LCOE) model. It is not perfect but it does allow technologies to be compared, broadly. Energy storage plants are different to generating plants in that the energy they use during their operations is electrical power from the grid rather than energy from the wind or coal. However once the differences are taken into account, it is feasible to estimate the levelized cost of the power from a storage plant. This figure can then be used to give some guidance regarding the situations in which it will be cost effective.

CAPITAL COSTS

Before considering levelized cost, it is worth looking at another important figure, the capital cost of a storage plant. This will be an important consideration when a storage facility is under consideration and may of itself influence the choice of technology. If one type of plant is much more expensive than another, then simple economics may dictate that the cheaper is built. There are two capital costs that can be important when looking as storage plants, the capital cost per unit of capacity usually measured in $/kW or similar and the capital cost per unit of storage, in $/kWh. Both are significant energy storage parameters.

The capital cost of storage plants per unit of capacity varies widely, even for the same technology. For example, the cost of pumped storage hydropower plants varies from $500/kW to $2500/kW. Smaller plants tend to be more expensive than larger plants. For compressed air energy storage (CAES) a typical cost estimate puts the cost at around $400/kW to $900/kW. Meanwhile a flywheel system can be expected to cost up to $2000/kW and that of super-capacitors is probably similar. This may seem relatively expensive but the flywheel or capacitor may be able to deliver its power much more quickly than alternative technologies.

Table 10.1[1] shows figures for the capital cost of energy storage in terms of cost per unit of energy stored. It is derived from a recent analysis by Lazard.[2] This table contains figures for six different storage technologies and for three different situations, transmission-grid-level storage, distribution-system storage at the substation-level, and "behind the meter" storage in industrial and commercial facilities.

[1]Lazard's Levelized Cost of Storage—Version 2.0, Lazard, 2016.
[2]Lazard's Levelized Cost of Storage—Version 2.0, Lazard, 2016.

Table 10.1 Capital Cost of Energy Storage

Technology	Transmission System Level ($/kWh)	Distribution Substation Level ($/kWh)	Industrial and Commercial ($/kWh)
Pumped storage hydropower	213–313	n/a	n/a
CAES	130–188	n/a	n/a
Lithium ion battery	386–917	432–901	452–1066
Vanadium flow battery	426–1026	631–1001	631–1001
Flywheel	n/a	551–949	551–949
Lead-acid battery	n/a	511–1211	551–1151
Source: Lazard.			

For large-scale transmission system storage the two primary contenders are pumped storage hydropower with an estimated cost of $213–313/kWh and CAES at $130–188/kWh. Lithium ion batteries ($386–917/kWh) and vanadium flow batteries ($426–1026/kWh) might also be used in small-scale grid installations.

At the distribution level neither pumped storage nor CAES is likely to be suitable because plant sizes are too large but other technologies are suitable. Four are shown in this table, lithium ion batteries with a cost of $432–901/kWh, lead-acid batteries ($511–1211/kWh), flywheels ($551–949$/kWh), and vanadium flow batteries ($631–1001/kWh). These same four technologies can also be used at the commercial and industrial level where they might be used to load shift or for power conditioning and support. Costs are broadly similar to those for distribution-level installations.

LEVELIZED COST OF ENERGY STORAGE

Similar figures for the levelized cost of energy storage are shown in Table 10.2.[3] Again there are six technologies and three different scenarios included. For transmission-grid-level storage, pumped storage hydropower has an estimated levelized cost of storage (LCOS) of $152–198/MWh while for CAES the estimated LCOS is $116–140/ MWh. Both offer more competitive storage than lithium ion batteries ($267–561/MWh) or vanadium flow batteries ($314–690/MWh). The

[3]Lazard's Levelized Cost of Storage—Version 2.0, Lazard, 2016.

Table 10.2 Levelized Cost of Energy Storage			
Technology	Transmission System Level ($/MWh)	Distribution Substation Level ($/MWh)	Industrial and Commercial ($/MWh)
Pumped storage hydropower	152–198	n/a	n/a
CAES	116–140	n/a	n/a
Lithium ion battery	267–561	345–657	530–1142
Vanadium flow battery	314–690	516–770	779–1164
Flywheel	n/a	400–654	623–1011
Lead-acid battery	n/a	425–933	648–1612
Source: Lazard.			

first two have consistently been considered the most cost effective for large-scale grid storage, load shifting, and load leveling.

At the distribution system level, the most competitive technology appears to be lithium ion batteries at $345–657/MWh followed by flywheels with an LCOS of $400–654/MWh, lead-acid batteries ($425–933/MWh), and vanadium flow batteries ($516–770/MWh). For industrial and commercial applications the order of competitiveness is the same but overall costs rise, reflecting the fact that smaller systems tend to be more expensive.

These figures can be put into perspective by looking at the cost of electricity to consumers. Recent figures from the US Energy Information Administration show that the average retail cost of electricity to residential consumers in October 2017 was as high as $213/MWh, depending on the state, and for industrial consumers it could be as high as $148/MWh.[4] Peak costs could be considerably higher. If the cost of energy from a storage plant is lower than the cost of power from the grid, then the storage plant offers a more cost-effective solution. These figures suggest that there are probably many situations where that is the case.

[4]Electric Power Monthly, October 2017, Table 5.6.A, US Energy Information Administration.

INDEX

Printed in the United States
By Bookmasters